Advanced Control Techniques for Power Converter and Drives

Advanced Control Techniques for Power Converter and Drives

Guest Editors

Daniele Scirè
Gianpaolo Vitale

MDPI

Basel • Beijing • Wuhan • Barcelona • Belgrade • Novi Sad • Cluj • Manchester

Guest Editors
Daniele Scirè
Department of Engineering
University of Palermo
Palermo
Italy

Gianpaolo Vitale
Institute for High
Performance Computing and
Networking
National Research Council
Palermo
Italy

Editorial Office
MDPI AG
Grosspeteranlage 5
4052 Basel, Switzerland

This is a reprint of the Special Issue, published open access by the journal *Electronics* (ISSN 2079-9292), freely accessible at: https://www.mdpi.com/journal/electronics/special_issues/P26FR87AS2.

For citation purposes, cite each article independently as indicated on the article page online and as indicated below:

Lastname, A.A.; Lastname, B.B. Article Title. *Journal Name* **Year**, *Volume Number*, Page Range.

ISBN 978-3-7258-5381-6 (Hbk)
ISBN 978-3-7258-5382-3 (PDF)
https://doi.org/10.3390/books978-3-7258-5382-3

Contents

About the Editors

Daniele Scirè

Daniele Scirè received an M.S. degree in Electronic Engineering and a Ph.D. degree in Information and Communication Technologies from the University of Palermo, Palermo, Italy, in 2017 and 2021, respectively. From 2018 to 2019, he was a visiting Ph.D. student with the Photovoltaic Material and Devices group, Delft University of Technology, the Netherlands, and in 2024 he was a visiting Researcher with the GREAH at Université de Normandie, France. Since 2022, he has been an Assistant Professor with the University of Palermo, Palermo, Italy. His current research interests include power electronics converters and their control, nonlinear magnetic components, modeling and design of solar cells, and thin-film deposition and characterization. Dr. Scirè was a recipient of the Graduate Student Award from the European Material Research Society in 2019 and of the Best Paper Award "Salvatore (Enzo) Piazza" from the Department of Engineering of the University of Palermo in 2022. Dr. Scirè is a reviewer for several journals and conferences.

Gianpaolo Vitale

Gianpaolo Vitale received an M.S. degree in electronic engineering from the University of Palermo, Palermo, Italy, in 1988. He is a Research Director of the Institute for High-Performance Computing and Networking, which is part of the National Research Council of Italy. He received his National Academic Qualification as a full professor in electronic engineering (cod 09_E3) in 2017 and his National Academic Qualification as a full professor in electric energy engineering (cod. 09/E2) in 2013. He teaches industrial electronics in the Department of Engineering, MD Electronics Engineering, University of Palermo. He is a Senior Member of IEEE. He has co-authored two books, three edited books, four book chapters, and more than a hundred scientific works. His current research interests include power electronics, power generation from renewable sources, electromagnetic compatibility, and robotics. He serves as a reviewer for several journals and conferences.

Preface

This Reprint on Advanced Control Techniques for Power Converters and Drives gathers a selection of recent research papers that address the growing complexity of power electronics and electric drive systems. The subject matter reflects the shift from conventional linear controllers toward predictive, adaptive, and passivity-based strategies, as well as advanced modulation methods that improve performance under stringent constraints.

The scope of this Reprint includes device-level modeling, real-time digital implementation, EMI-aware control, and distributed coordination of converters in large-scale energy and transportation systems. Contributions highlight both theoretical developments and experimental validations, showing how advanced control can be practically realized in motor drives, renewable energy integration, multiport architectures, and emerging aerospace and industrial applications.

The purpose of this collection is to provide a comprehensive reference for the community, showcasing how new control paradigms can enhance efficiency, robustness, and resilience while addressing challenges such as nonlinear component behavior, thermal effects, and system-level uncertainty.

The motivation for preparing this Reprint stems from the rapid adoption of wide-bandgap devices, the increasing penetration of converter-dominated grids, and the demand for certifiable, high-performance control strategies in safety-critical environments.

This Reprint is addressed to researchers, graduate students, and practicing engineers in the fields of power electronics, control systems, and electrical drives. It also serves as a resource for professionals working in renewable energy, electric mobility, aerospace, and industrial automation who are seeking insights into the future of advanced control solutions for power converters and drives.

Daniele Scirè and Gianpaolo Vitale
Guest Editors

electronics MDPI

Editorial

Advanced Control Techniques for Power Converter and Drives

Daniele Scirè [1,*] and Gianpaolo Vitale [2,*]

[1] Department of Engineering, University of Palermo, 90128 Palermo, Italy
[2] Consiglio Nazionale delle Ricerche, ICAR, Institute for High Performance Computing and Networking, Italian National Research Council of Italy, 90146 Palermo, Italy
* Correspondence: daniele.scire@unipa.it (D.S.); gianpaolo.vitale@icar.cnr.it (G.V.)

1. Introduction

Advanced control is now central to performance, robustness, and the ability to achieve the highest power density in modern power converters. Beyond classical linear designs, research has rapidly expanded across model predictive control, adaptive and robust methods, passivity-based strategies, and learning-enabled controllers, together with spectrum-shaping modulation to address the EMI and vibro-acoustics in drive systems [1–10]. These techniques are being exploited in demanding contexts such as grid-forming inverters, hybrid AC/DC networks, and electrified transportation, where tight transience, device stress limits, and grid codes must all be satisfied simultaneously.

Despite the progress, several gaps remain regarding the control design, device physics, and implementation: (i) accurate handling of nonlinear magnetic components within real-time controllers; (ii) stability guarantees for distributed and parallelized converter architectures; (iii) realizations of complex control schemes on embedded platforms (e.g., FPGAs); and (iv) grid-interactive scheduling and optimization frameworks that explicitly treat uncertainty while coordinating power–electronic elements in distribution networks.

Recent work has started to close these gaps. For example, the authors of [11] propose a quasi-constant on-time (QCOT) control for SMPS operating with nonlinear temperature-dependent inductors; by estimating the power switch conduction time and exploiting the saturation safely, the QCOT raises the inductor current capability and power density while avoiding thermal runaway. In [12], a 4 MW high-power-density generator for hybrid-electric aircraft, targeting gravimetric power densities around 20 kW/kg with advanced PM design and thermal management is presented. In [13], the authors propose a multiport power conversion system (MPCS) for the More Electric Aircraft, enabling fault-tolerant ring power distribution with minimal weight penalty. An advanced discontinuous PWM for multilevel cascaded H-bridge converters, reducing switching losses while mitigating harmonic degradation in N-cell structures, is reported in [14]. A modulated model-predictive integral control for synchronous reluctance motor drives, ensuring fixed switching frequency, low ripple, and robustness against parameter mismatches is presented in [15]. Finally, ref. [16] explores sampling-time harmonic control for cascaded H-bridges under active thermal control, addressing lifetime extension while suppressing low-order distortion.

This Special Issue was conceived to collect the latest research across advanced control theory, power-device and passive modeling, and embedded implementation and to highlight solutions that translate into experimentally validated performance gains in the specific field of power electronics.

In order to meet the demand for new contributions also relating to the abovementioned topics, we are pleased to announce that a Second Edition of this Special Issue is now open

 https://doi.org/10.3390/electronics14183710

for submissions. We particularly welcome innovative contributions to the field of advanced control techniques for power electronics converters and drives. For further details, please see https://www.mdpi.com/journal/electronics/special_issues/M485MV576M, (accessed on 15 September 2025).

2. Highlighting Key Contributions

This Special Issue brings together a selection of innovative research articles that demonstrate the state of the art in control methodologies for power converters and electric drives. The following contributions reflect significant progress in areas including harmonic mitigation, voltage stability, efficiency optimization, and intelligent control implementation.

2.1. Double-Loop Controller Design of a Single-Phase 3-Level Power Factor Correction Converter

Han and Kim (Contribution 1) detail a practical inner–outer loop design (SISOTOOL-based) augmented with a targeted 120 Hz band-stop to suppress ripple injection while preserving the dynamic response. The hardware results show clean boost and buck transitions (210 V to 150 V) without overshoot and stable behavior under load steps (50 Ω to 25 Ω), matching the PSIM predictions.

2.2. Robust PI-PD Controller Design: Industrial Simulation Case Studies and a Real-Time Application

Alyoussef, Kaya, and Akrad (Contribution 2) propose a geometry-driven robust design method that characterizes the controller parameter region, guaranteeing closed-loop stability and adequate margins, which then selects a PI–PD operating point near the region centroid. Hardware-in-the-loop and real-time tests on a twin-rotor MIMO system (TRMS) confirm the setpoint tracking and disturbance rejection with low tuning effort and transparent robustness guarantees.

2.3. Advanced Distributed Control of Parallel Resonant CLLC DAB Converters

Vicente et al. (Contribution 3) propose a scalable distributed architecture that combines a master voltage controller with local current controllers to balance load and suppress circulating currents in parallel CLLC DAB stages. The experiments demonstrated stability, an ∼80% faster transient via a feed-forward path, and a current-sharing deviation <3% from light to full load, achieved without a fragile centralized current bus, making it attractive for hybrid AC/DC microgrids and SST front-ends.

2.4. FPGA Implementation of Nonlinear Model Predictive Control for a Boost Converter with a Partially Saturating Inductor

Ravera et al. (Contribution 4) embed a nonlinear inductor and thermal-aware converter model into an NMPC solved by Mesh Adaptive Direct Search, mapped to an AMD/Xilinx FPGA. The co-simulation and experiments show sub-millisecond voltage regulation under steps, while respecting current constraints; practical figures include a control latency of ∼16.6 μs and operation up to 60 kHz sampling, illustrating a viable path to high-speed certifiable MPC in power supplies.

2.5. Dual-Random Space Vector Pulse Width Modulation Strategy Based on Optimized Beta Distribution

Gu et al. (Contribution 5) introduce a work to reduce high-frequency current harmonics and associated acoustic/vibration signatures in PMSMs; the work randomizes both the switching frequency and zero-vector selection. Using a PSO-tuned Beta distribution for the RNG yielded superior spectral spreading versus classical LCG methods; motor-bench experiments validated the reductions in high-frequency vibration while preserving the dynamic performance.

2.6. *Stochastic Operation of BESS and MVDC Link in Distribution Networks Under Uncertainty*

Han, Song, and Lee (Contribution 6) introduce a distributionally robust chance-constrained (DRCC) scheduler that coordinates MVDC link setpoints and BESS dispatch across interconnected feeders with PV/load uncertainty. Case studies quantify the cost–reliability trade, e.g., up to 44.7% operational cost reduction, while maintaining ≈96.8% bus-voltage reliability, showing how probabilistic reliability constraints can be tuned to system economics.

2.7. *A Passivity-Based Control Integrated with Virtual DC Motor Strategy for Boost Converters Feeding Constant Power Loads*

Ou et al. (Contribution 7) present a passivity observer/controller framework with a "virtual DC motor" current loop that yields robust damping and constraint handling for grid-forming inverters under grid disturbances. The method maintains synchronization and current limits without retuning across operating points, and the experiments indicated improved stability margins relative to standard inner loops.

2.8. *Symmetric Optimization Strategy Based on Triple-Phase Shift for Dual-Active Bridge Converters with Low RMS Current and Full ZVS over Ultra-Wide Voltage and Load Ranges*

Cui et al. (Contribution 8) focus on RMS-current minimization and ZVS extension; this paper synthesizes recent DAB modulation strategies and outlines implementable heuristics for efficiency over wide operating ranges. It provides design guidance that complements distributed and resonant topologies elsewhere in this Special Issue.

Funding: This research received no external funding.

Acknowledgments: We extend our appreciation to all the authors who contributed their research to this Special Issue with their innovative work. We are grateful to the reviewers for their thorough and insightful evaluations, which greatly enhanced the scientific rigor and clarity of the published articles. Our special thanks also go to the editorial team and staff of *Electronics* at MDPI for their continued support and for providing us with the opportunity to serve as Guest Editors of this Special Issue.

Conflicts of Interest: The authors declare no conflicts of interest.

List of Contributions

1. Han, J.-H.; Kim, I.-S. Double-Loop Controller Design of a Single-Phase 3-Level Power Factor Correction Converter. *Electronics* **2024**, *13*, 2863. https://doi.org/10.3390/electronics13142863.
2. Alyoussef, F.; Kaya, I.; Akrad, A. Robust PI-PD Controller Design: Industrial Simulation Case Studies and a Real-Time Application. *Electronics* **2024**, *13*, 3362. https://doi.org/10.3390/electronics13173362.
3. Vicente, D.C.; Carrero, A.M.; Díez, E.G.; Solís, J.M.C.; Rubio, F.R. Advanced Distributed Control of Parallel Resonant CLLC DAB Converters. *Electronics* **2025**, *14*, 318. https://doi.org/10.3390/electronics14020318.
4. Ravera, A.; Oliveri, A.; Lodi, M.; Storace, M. FPGA Implementation of Nonlinear Model Predictive Control for a Boost Converter with a Partially Saturating Inductor. *Electronics* **2025**, *14*, 941. https://doi.org/10.3390/electronics14050941.
5. Gu, X.; Wu, K.; Jin, X.; Zhang, G.; Chen, W.; Li, C. Dual-Random Space Vector Pulse Width Modulation Strategy Based on Optimized Beta Distribution. *Electronics* **2025**, *14*, 1779. https://doi.org/10.3390/electronics14091779.
6. Han, C.; Song, S.; Lee, J. Stochastic Operation of BESS and MVDC Link in Distribution Networks Under Uncertainty. *Electronics* **2025**, *14*, 2737. https://doi.org/10.3390/electronics14132737.

7. Ou, M.; Gong, P.; Guo, H.; Li, G. A Passivity-Based Control Integrated with Virtual DC Motor Strategy for Boost Converters Feeding Constant Power Loads. *Electronics* **2025**, *14*, 2909. https://doi.org/10.3390/electronics14142909.
8. Cui, L.; Zhang, Y.; Wang, X.; Zhang, D. Symmetric Optimization Strategy Based on Triple-Phase Shift for Dual-Active Bridge Converters with Low RMS Current and Full ZVS over Ultra-Wide Voltage and Load Ranges. *Electronics* **2025**, *14*, 3031. https://doi.org/10.3390/electronics14153031.

References

1. Schwenzer, M.; Ay, M.; Bergs, T.; Abel, D. Review on model predictive control: An engineering perspective. *Int. J. Adv. Manuf. Technol.* **2021**, *117*, 1327–1349. [CrossRef]
2. Panda, S.K.; Subudhi, B. A Review on Robust and Adaptive Control Schemes for Microgrid. *J. Mod. Power Syst. Clean Energy* **2023**, *11*, 1027–1040. [CrossRef]
3. Gholami-Khesht, H.; Davari, P.; Blaabjerg, F. Chapter 5—Adaptive control in power electronic systems. In *Control of Power Electronic Converters and Systems*; Blaabjerg, F., Ed.; Academic Press: Cambridge, MA, USA, 2021; pp. 125–147. [CrossRef]
4. Ren, Q.; Chen, A.; Fang, J.; Liu, X.; Liu, T.; Zhang, G.; Zhang, C. An Improved Passivity-Based Direct Power Control Strategy for AC/DC Converter Under Unbalanced and Distorted Grid Conditions. *IEEE Trans. Power Electron.* **2023**, *38*, 12153–12165. [CrossRef]
5. Khalid, M. Passivity-Based Nonlinear Control Approach for Efficient Energy Management in Fuel Cell Hybrid Electric Vehicles. *IEEE Access* **2024**, *12*, 84169–84188. [CrossRef]
6. Qiu, Z.; Chen, Y.; Liu, X.; Kang, Y.; Liu, H. Analysis of the sideband current harmonics and vibro-acoustics in the PMSM with SVPWM. *IET Power Electron.* **2020**, *13*, 1033–1040. [CrossRef]
7. Ibrahim, A.; Salem, M.; Kamarol, M.; Delgado, M.T.; Desa, M.K.M. Review of Active Thermal Control for Power Electronics: Potentials, Limitations, and Future Trends. *IEEE Open J. Power Electron.* **2024**, *5*, 414–435. [CrossRef]
8. Yu, Z.; Long, J. Review on Advanced Model Predictive Control Technologies for High-Power Converters and Industrial Drives. *Electronics* **2024**, *13*, 4969. [CrossRef]
9. Yusuf, S.S.; Kunya, A.B.; Abubakar, A.S.; Salisu, S. Review of load frequency control in modern power systems: A state-of-the-art review and future trends. *Electr. Eng.* **2025**, *107*, 5823–5848. [CrossRef]
10. Zelba, M.; Deveikis, T.; Gudžius, S.; Jonaitis, A.; Bandza, A. Review of Power Control Methods for a Variable Average Power Load Model Designed for a Microgrid with Non-Controllable Renewable Energy Sources. *Sustainability* **2023**, *15*, 9100. [CrossRef]
11. Scirè, D.; Lullo, G.; Vitale, G. A Quasi-Constant On-Time Control for SMPS with a Nonlinear Inductor Based on Power Switch Conduction Time Estimation. *IEEE Trans. Ind. Inform.* **2024**, *21*, 1419–1428. [CrossRef]
12. Golovanov, D.; Gerada, D.; Sala, G.; Degano, M.; Trentin, A.; Connor, P.H.; Xu, Z.; Rocca, A.L.; Galassini, A.; Tarisciotti, L.; et al. 4-MW Class High-Power-Density Generator for Future Hybrid-Electric Aircraft. *IEEE Trans. Transp. Electrif.* **2021**, *7*, 2952–2964. [CrossRef]
13. Gu, C.; Yan, H.; Yang, J.; Sala, G.; De Gaetano, D.; Wang, X.; Galassini, A.; Degano, M.; Zhang, X.; Buticchi, G. A Multiport Power Conversion System for the More Electric Aircraft. *IEEE Trans. Transp. Electrif.* **2020**, *6*, 1707–1720. [CrossRef]
14. Marquez, A.; Monopoli, V.G.; Tcai, A.; Leon, J.I.; Buticchi, G.; Vazquez, S.; Liserre, M.; Franquelo, L.G. Discontinuous-PWM Method for Multilevel *N*-Cell Cascaded H-Bridge Converters. *IEEE Trans. Ind. Electron.* **2021**, *68*, 7996–8005. [CrossRef]
15. Riccio, J.; Karamanakos, P.; Odhano, S.; Tang, M.; Nardo, M.D.; Tresca, G.; Zanchetta, P. Modulated Model-Predictive Integral Control Applied to a Synchronous Reluctance Motor Drive. *IEEE J. Emerg. Sel. Top. Power Electron.* **2023**, *11*, 3000–3010. [CrossRef]
16. Marquez, A.; Monopoli, V.G.; Leon, J.I.; Ko, Y.; Buticchi, G.; Vazquez, S.; Liserre, M.; Franquelo, L.G. Sampling-Time Harmonic Control for Cascaded H-Bridge Converters with Thermal Control. *IEEE Trans. Ind. Electron.* **2020**, *67*, 2776–2785. [CrossRef]

electronics

MDPI

Article

Double-Loop Controller Design of a Single-Phase 3-Level Power Factor Correction Converter

Jun-Hyuk Han and IL-Song Kim *

Department of Electrical Engineering, Korea National University of Transportation, Chungju 27469, Republic of Korea; 22028405@a.ut.ac.kr
* Correspondence: iskim@ut.ac.kr

Abstract: This paper presents a study on the double-loop controller design technique for a single-phase power factor correction (PFC) three-level (TL) boost converter. Designing a double-loop controller using conventional methods is challenging due to the 120 Hz voltage ripple in the output voltage. This study proposes new double-loop control design methods using a band-stop filter and MATLAB SISOTOOL, detailed in a step-by-step sequence. A band-stop filter with a 120 Hz stop band is applied to the double-loop controller design. Modeling based on the state-space equation, applicable to the full duty range, is constructed to obtain the transfer function. The double-loop controller structure is then designed, and optimal gains that satisfy the design requirements are obtained through automatic tuning using the MATLAB SISOTOOL library. The simulation results demonstrate the performance of the proposed method, which is further verified by experimental results.

Keywords: three-level boost converter; PFC; double-loop control; MATLAB SISOTOOL

1. Introduction

Global population growth, economic growth, industrialization, and technological advancements continue to increase the demand for today's limited energy resources. Electronic systems, in particular, account for a significant portion of energy consumption. Efficient management and conservation of these systems can significantly contribute to energy savings and cost reduction. In response to the growing demand for limited energy resources and rising energy prices, improving power efficiency in electronic systems has become a critical challenge.

In recent years, research on efficient energy conversion technologies has been active, and power factor correction (PFC) has garnered significant attention, particularly for single-phase PFC. Power factor correction offers several benefits, including improved power quality, increased energy efficiency, and reduced power losses.

The circuit to implement a single-phase PFC consists of a voltage controller to regulate the direct current (DC) voltage at the output and a current controller to control the alternating current (AC) at the input, maintaining it in a sinusoidal wave to achieve a nearly uniform power factor. The detailed configuration is presented in Section 3, "Double Loop Controller Design". In a single-phase PFC system, there is only one power source. Therefore, the single-phase AC is passed through an all-pass filter (APF) to create a virtual AC with a 90-degree phase shift. This virtual AC is then converted to DC through a d-q conversion for control purposes. With d-q conversion, alternating current is effectively converted to direct current, allowing for control and regulation in a single-phase PFC system [1]. This approach is widely used due to its high efficiency, simple topology, continuous input current, and step-up voltage conversion ratio. Another approach could involve using the instantaneous AC control method; however, this approach may introduce steady-state errors.

Single-phase PFC systems often utilize boost converters to leverage these benefits. Boost converters are preferred in single-phase PFCs because they can adjust the output voltage by boosting it while maintaining a high power factor.

However, with the recent increase in the use of high-power applications, the use of single-level boost converters in PFC circuits has increased the volume and weight of the inductor, and losses in the power devices have become a major factor affecting the cost and efficiency of the converter. These applications are energy-intensive and consume significant power from the power network. High-power applications, such as electric vehicles, electric buses, railroad cars, power plants, substations, and distribution systems in energy production, as well as distribution systems for electric vehicles, require substantial power.

To compensate for the disadvantages of the single boost converter structure, the application of a boost converter with a three-level (TL) structure to PFC has been studied as a possible solution [2–5]. Compared to conventional boost converters, TL boost converters offer higher efficiency and a wider control range. However, since single-phase PFC circuits generate a 120 Hz ripple at the output, designing a double-loop controller is challenging. This challenge is further compounded when adopting a TL boost converter structure. Previous research has utilized notch filters or lead-lag compensators to compensate for the 120 Hz ripple. However, these methods can potentially induce instability and complicate the design of a double-loop controller [6].

This paper presents the results of a study in which a single-phase PFC double-loop controller design technique, accommodating the 120 Hz ripple, is applied to a TL boost converter. After constructing the state-space equations applicable to the entire duty range to obtain the transfer function, a double-loop controller was designed using MATLAB SISOTOOL. The validity of the designed controller was verified through PSIM simulations and experimental results.

2. System Modeling

The circuit diagram of the single-phase three-level (TL) PFC converter with a diode bridge rectifier is shown in Figure 1. This configuration has the disadvantage of an increased number of semiconductor switches, diodes, and capacitors due to the cascade connection compared to the two-level configuration. However, it also has advantages, such as a reduced volume and weight of the inductor and reduced voltage stress on the semiconductor switches. In fact, the three-level configuration can reduce the voltage ratings of the semiconductor switches and diodes to half of those in the two-level configuration. Therefore, the three-level PFC converter can significantly reduce material costs.

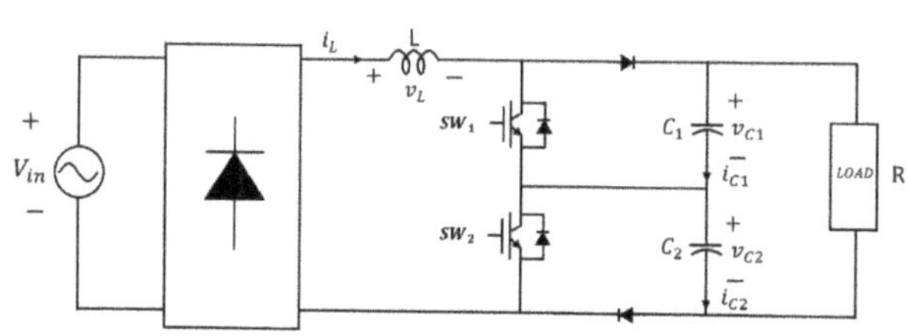

Figure 1. The circuit diagram of TL PFC converter.

The dynamic mathematical model of the TL PFC converter can be obtained by analyzing the switch operations. The equivalent circuit of the TL PFC converter is divided into four operating states according to the switching operation, as shown in Figure 2. The state when both switches, SW_1 and SW_2, are off is shown in Figure 2a. The state when SW_1 is on while SW_2 is off is shown in Figure 2b. The state when SW_1 is off, and SW_2 is on is shown in Figure 2c. Finally, Figure 2d depicts the state when both switches, SW_1 and SW_2, are on.

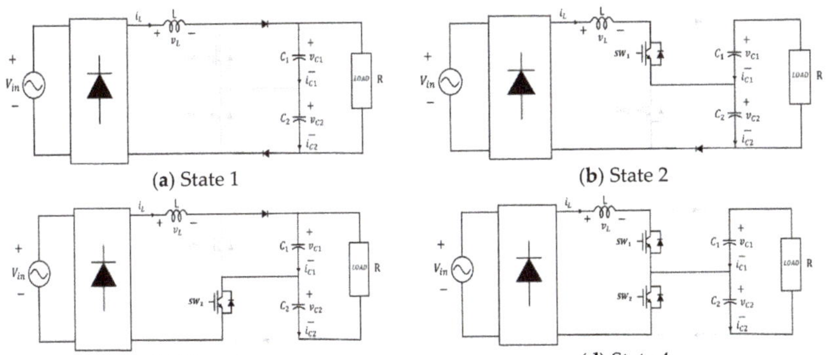

(**a**) State 1 (**b**) State 2 (**c**) State 3 (**d**) State 4

Figure 2. TL PFC converter switching operation circuit.

The average state space equation can be derived by analyzing the operation mode determined by the duty boundary of 0.5. When the duty cycle is less than 0.5, it is defined as Mode 1. The operating sequence for Mode 1 is State 1 → State 2 → State 1 → State 3, as depicted in Figure 3a. Conversely, when the duty cycle is greater than 0.5, it is defined as Mode 2. The operating sequence for Mode 2 is State 4 → State 3 → State 4 → State 2, as shown in Figure 3b. Since the system control involves two operation modes with a duty boundary of 0.5, it is necessary to analyze their operations for both modes. The state space equations for each state are summarized in Table 1.

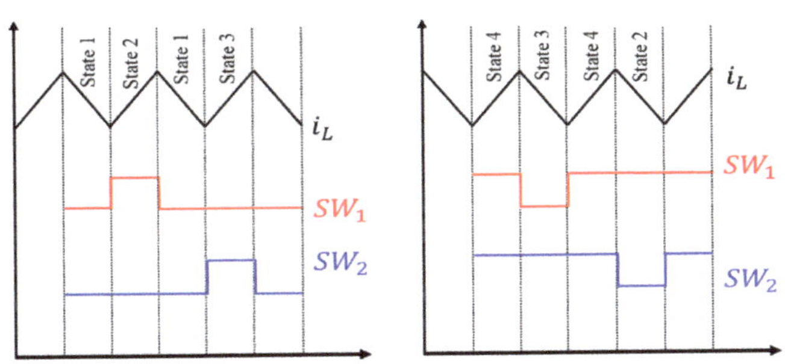

(**a**) Mode 1 [duty < 0.5] (**b**) Mode 2 [duty > 0.5]

Figure 3. Mode operation by duty range.

Table 1. The summary of the state space equation for each operation states.

State 1: [SW_1, SW_2 OFF] Operation time : $(1-2d)Ts$ $\dot{i}_L = \frac{v_{in}}{L} - \frac{(v_{C1}+v_{C2})}{L}$ $\dot{v}_{c1} = \frac{i_L}{C_1} - \frac{(v_{C1}+v_{C2})}{RC_1}$ $\dot{v}_{c2} = \frac{i_L}{C_2} - \frac{(v_{C1}+v_{C2})}{RC_2}$ (1)	State 2: [SW_1 ON, SW_2 OFF] Operation time : $(d)Ts$ $\dot{i}_L = \frac{v_{in}}{L} - \frac{v_{C2}}{L}$ $\dot{v}_{c1} = -\frac{(v_{C1}+v_{C2})}{RC_1}$ $\dot{v}_{c2} = \frac{i_L}{C_2} - \frac{(v_{C1}+v_{C2})}{RC_2}$ (2)
State 3: [SW_1 OFF, SW_2 ON] Operation time : $(1-d)Ts$ $\dot{i}_L = \frac{v_{in}}{L} - \frac{v_{C1}}{L}$ $\dot{v}_{c1} = \frac{i_L}{C_1} - \frac{(v_{C1}+v_{C2})}{RC_1}$ $\dot{v}_{c2} = -\frac{(v_{C1}+v_{C2})}{RC_2}$ (3)	State 4: [SW_1, SW_2 ON] Operation time : $(2d-1)Ts$ $\dot{i}_L = \frac{v_{in}}{L}$ $\dot{v}_{c1} = -\frac{(v_{C1}+v_{C2})}{RC_1}$ $\dot{v}_{c2} = -\frac{(v_{C1}+v_{C2})}{RC_2}$ (4)

The state-averaging equations can be calculated by multiplying the state equations and their operation time for each Mode. The resultant state-averaging equations are the same for Mode 1 and Mode 2 and can be expressed as displayed in (5).

$$\begin{pmatrix} \dot{i}_L \\ \dot{v}_{c1} \\ \dot{v}_{c2} \end{pmatrix} = \begin{pmatrix} 0 & -\frac{(1-d)}{L} & -\frac{(1-d)}{L} \\ -\frac{(1-d)}{C_1} & -\frac{1}{RC_1} & -\frac{1}{RC_1} \\ -\frac{(1-d)}{C_2} & -\frac{1}{RC_2} & -\frac{1}{RC_2} \end{pmatrix} \begin{pmatrix} i_L \\ v_{C1} \\ v_{C2} \end{pmatrix} + \begin{pmatrix} \frac{1}{L} \\ 0 \\ 0 \end{pmatrix} v_{in} \tag{5}$$

The state-averaging equation of Mode 1 is the same as that of Mode 2, which means that mode transition does not need to be considered when designing the controller. This means that the above equation can be applicable to the full duty range. However, since the above equation includes duty, that is, a variable that changes with time, linearization at the operating point is necessary [7–11]. By applying small signal perturbation for linearization at the operating point, the following linearized equation can be obtained:

$$\begin{pmatrix} \hat{i}_L \\ \hat{v}_{c1} \\ \hat{v}_{c2} \end{pmatrix} = \begin{pmatrix} 0 & -\frac{(1-D)}{L} & -\frac{(1-D)}{L} \\ \frac{(1-D)}{C_1} & -\frac{1}{RC_1} & -\frac{1}{RC_1} \\ \frac{(1-D)}{C_2} & -\frac{1}{RC_2} & -\frac{1}{RC_2} \end{pmatrix} \begin{pmatrix} \hat{i}_L \\ \hat{v}_{c1} \\ \hat{v}_{c2} \end{pmatrix} + \begin{pmatrix} \frac{(V_{c1}+V_{c2})}{L} \\ -\frac{I_L}{C_1} \\ -\frac{I_L}{C_2} \end{pmatrix} \hat{d} \tag{6}$$

If assuming that capacitor C_1 and C_2 are equal in the small signal analysis equation, it can be expressed as a quadratic equation, and, thus, a simple controller design is possible. The capacitor imbalance occurs in actual applications. Therefore, the capacitor voltage balancing controller is required for the reduced quadratic equation. The balancing controller design is shown in the controller design section. The final linearized state space equation is shown as follows:

$$\begin{pmatrix} \dot{i}_L \\ \dot{v}_{c0} \end{pmatrix} = \begin{pmatrix} 0 & -\frac{(1-D)}{L} \\ \frac{(1-D)}{C_0} & -\frac{1}{RC_0} \end{pmatrix} \begin{pmatrix} i_L \\ v_{C0} \end{pmatrix} + \begin{pmatrix} \frac{1}{L} \\ 0 \end{pmatrix} v_{in} \tag{7}$$

And the derived steady-state DC values of V_{C0}, I_L are shown in (8).

$$I_L = \frac{V_{in}}{R(1-D)}, V_{C0} = \frac{V_{in}}{(1-D)} \tag{8}$$

The input–output transfer function equation for voltage and current controller design can be obtained from the Laplace transform and is shown in the following equations from (9) to (11):

$$\frac{\hat{i}_L(s)}{\hat{d}(s)} = G_1(s) = \frac{\frac{V_{C0}}{L}s + \frac{V_{C0}}{RC_0L} + \frac{I_L}{LC_0}(1-D)}{s^2 + \frac{1}{RC_0} + \frac{1}{LC_0}(1-D)^2} \tag{9}$$

$$\frac{\hat{v}_{co}(s)}{\hat{d}(s)} = G_2(s) = \frac{-\frac{I_L}{C_0}s + \frac{V_{C0}}{LC_0}(1-D)}{s^2 + \frac{1}{RC_0} + \frac{1}{LC_0}(1-D)^2} \tag{10}$$

$$\frac{\hat{v}_{c0}(s)}{\hat{i}_L(s)} = G_3(s) = \frac{-\frac{I_L}{C_0}s + \frac{V_{C0}}{LC_0}(1-D)}{\frac{V_{C0}}{L}s + \frac{V_{C0}}{RC_0L} + \frac{I_L}{LC_0}(1-D)} \tag{11}$$

3. Double-Loop Controller Design

The configuration of the single-phase PFC controller consists of a double-loop structure, as illustrated in Figure 4. It comprises a voltage controller for regulating the DC output voltage in the outer loop and a current controller for ensuring the power factor remains close to unity by controlling the current of the AC input terminal as a sine wave in the inner loop. In single-phase PFC control, the Proportional-Integral (PI) controller cannot be directly applied to the current controller because the current waveform is sinusoidal, not DC. Conventional methods often use an instantaneous AC current controller that adopts a Proportional (P) controller. However, these methods may fail to achieve zero steady-state error and can be problematic when there are abrupt changes in the command or disturbances in the environment.

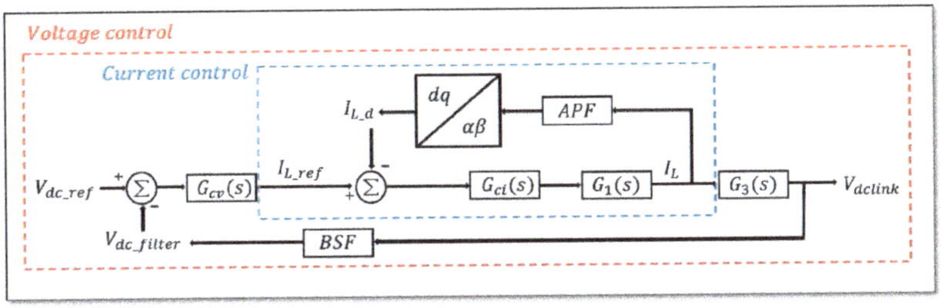

Figure 4. Double-loop controller configuration.

To achieve zero steady-state error, a PI current controller using DQ transformation can be applied. However, this transformation typically requires at least two sources, which poses a challenge in single-source systems. To overcome this limitation, a virtual 90-degree phase-shifted signal is created. This can be achieved by using a virtual power source or implementing a filter such as an All Pass Filter (APF) [12]. This paper explores the use of APF to facilitate single-phase DQ conversion.

The detailed control block diagram of the outer loop, regulating the output voltage, and the inner loop, controlling the inductor current, is depicted in Figure 4. The output of the voltage controller, I_{L_ref} (inductor current reference), is generated through voltage control at the DC output stage. However, in single-phase PFC, voltage ripple occurs at twice the AC input power frequency and manifests in the DC link voltage [13,14]. This ripple not only complicates voltage control but also impacts current control, posing challenges in designing single-phase PFC controllers and determining PI gain values. To address this issue, this paper proposes the use of a band-stop filter in voltage control to suppress the 120 Hz DC voltage ripple. The equation for the Band-stop filter is expressed in (12).

$$BSF = \frac{s^2 + {\omega_0}^2}{s^2 + Bs + {\omega_0}^2} \tag{12}$$

From (12), the equations for the stop-band frequency, B and ω_0, can be obtained through (13) and (14), respectively. f_b represents the stop-band frequency, and f_c represents the reference frequency [15].

$$f_b = \frac{B}{2\pi} \rightarrow B = 2\pi f_b \tag{13}$$

$$f_C = \frac{\omega_0}{2\pi} \rightarrow \omega_0 = 2\pi f_c \tag{14}$$

By using (13) and (14), the values corresponding to each frequency can be calculated. The transformation from the s-domain (continuous) to the z-domain (discrete), is shown in (15) which is derived by substituting s with $\frac{2}{T}\frac{1-z^{-1}}{1+z^{-1}}$ in the s-domain equation and expanding it.

$$BSF(Z) = \frac{\left(\frac{4}{T^2} + {\omega_0}^2\right) + \left(-\frac{8}{T^2} + 2{\omega_0}^2\right)z^{-1} + \left(\frac{4}{T^2} - \frac{2\omega_0}{2T} + {\omega_0}^2\right)z^{-2}}{\left(\frac{4}{T^2} + \frac{2\omega_0}{2T} + \omega_0\right) + \left(-\frac{8}{T^2} + 2{\omega_0}^2\right)z^{-1} + \left(\frac{4}{T^2} - \frac{2\omega_0}{2T} + {\omega_0}^2\right)z^{-2}} \tag{15}$$

The 120 Hz ripple can be effectively removed using a band-stop filter, as described by Equation (15). The selected parameters are f_b = 9.55 [Hz] and f_C = 120 [Hz]. Equation (15) has been implemented on a microcontroller, such as a Digital Signal Processor (DSP), using C-code programming. The filter can be applied to both 50 Hz and 60 Hz frequencies by adjusting the parameters of the band-stop filter (BSF). While the manuscript focuses on 60 Hz operation due to its relevance to the grid system used, the flexibility to accommodate both frequencies is noted. Adjusting the BSF parameters enables straightforward adaptation to 50/60 Hz operation.

For the design of the voltage controller, specifications are set with a phase margin of 90 degrees and a bandwidth of 50 [rad/s]. The optimal gains that satisfy these design specifications are automatically determined using the MATLAB SISOTOOL library. The transfer function of the final designed voltage controller can be expressed as shown in Equation (16).

$$G_{cv}(s) = 0.2015 \times \frac{s+16.2}{s} \tag{16}$$

The resultant Bode plot for the designed voltage controller is presented in Figure 5. It can be observed that the phase margin is 90.1 degrees, and the bandwidth is 50.2 [rad/s], meeting the specified design criteria. Therefore, the stability of the system can be confirmed through the Bode plot analysis.

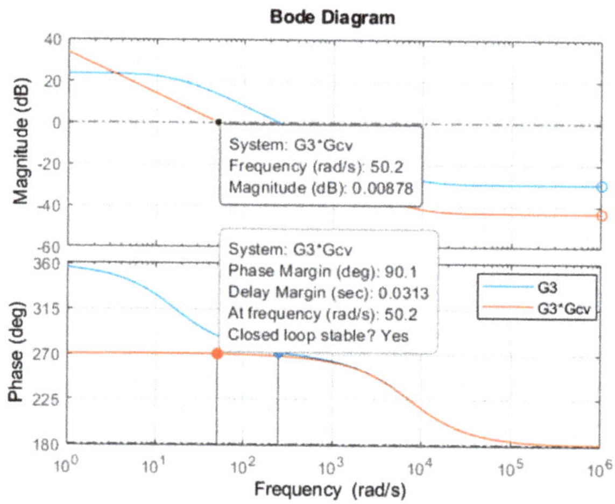

Figure 5. TL PFC converter voltage controller Bode plot.

In the current control configuration, an APF is used to create a virtual 90-degree phase axis, enabling the design of a single-phase DQ transformation. The APF used in the current controller is a mathematical equation that introduces a phase delay without altering the magnitude of the input signal. It became a transfer function with a magnitude of one across all frequency bands and a phase delay ranging from 0 to 180 deg. (1-pole filter). By utilizing the APF, the actual current axis I_α and the virtual current axis I_β were generated, as expressed in Equation (17) [16,17]. The DQ transformation formula based on (17) is described in (18).

In the current control configuration, an All Pass Filter (APF) is employed to generate a virtual 90-degree phase axis, facilitating the design of a single-phase DQ transformation. The APF utilized in the current controller is a mathematical equation that introduces a phase delay while preserving the magnitude of the input signal. It operates as a transfer function with a magnitude of one across all frequency bands and introduces a phase delay ranging from 0 to 180 degrees (a 1-pole filter). By utilizing the APF, both the actual current axis I_α and the virtual current axis I_β are generated, as expressed in Equation (17). The DQ transformation formula based on Equation (17) is described in Equation (18).

$$I_\alpha(k) = I_L$$

$$I_\beta(k) = 0.9813 \cdot I_\beta(k-1) + 0.9813 I_L(k) - I_L(k-1) \tag{17}$$

$$\begin{bmatrix} I_{Ld} \\ I_{Lq} \end{bmatrix} = \begin{bmatrix} cos\theta & sin\theta \\ -sin\theta & cos\theta \end{bmatrix} \begin{bmatrix} I_{\alpha} \\ I_{\beta} \end{bmatrix} \tag{18}$$

Current control is achieved through the use of an APF and DQ transformation. The design specifications for the current controller are set with a phase margin of 60 degrees and a bandwidth of 4800 [rad/s]. The transfer function of the designed current controller is expressed in Equation (19).

$$G_{ci} = 0.019522 \times \frac{s + 2690}{s} \tag{19}$$

The Bode plot for the current controller is presented in Figure 6. It can be observed from the plot that the phase margin is 60.8 degrees, and the bandwidth is 4820 [rad/s], meeting the specified design criteria. Therefore, the stability of the system can be confirmed through the Bode plot analysis.

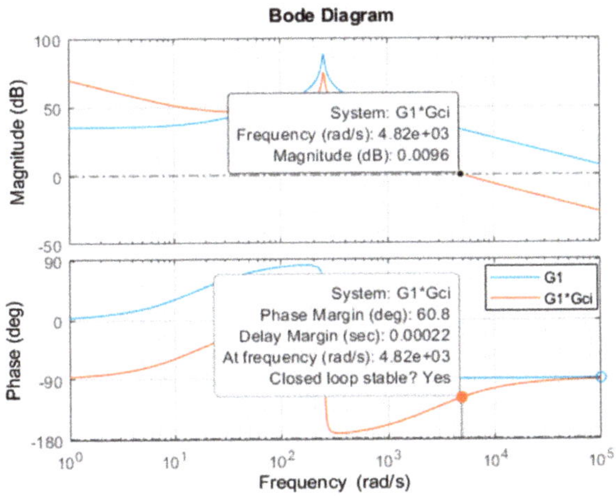

Figure 6. TL PFC converter current controller Bode plot.

The MATLAB m-file, which was used in the controller design, is shown below:

```
s = tf('s');

Vin = 310;
L = 2.4 × 10^-3;   % L = 2.4 mh
C = 2400 × 10^-6;  %C = 2400 μf
R = 50;
D = 0.311;
D1 = 1 − D;
Vc = Vin/D1;
IL = Vc/(R*D1);

G1 = (Vc/L)*(s + 1/(R*C) + (1 − D)*IL/(Vc*C))/(s^2 + s/(R*C)+(1 − D)^2/(L*C));
G2 = −Vin/(L*C)*(−1 + L*s/(R*(1 − D)^2))/(s^2 + s/(R*C) + (1 − D)^2/(L*C));
G3 = (IL/Vc)*(L/C)*(−s + (1-D)*Vc/(L*IL))/(s + 1/(R*C) + (1 − D)*IL/(Vc*C));

Kpv = 1; Kiv = 1;        % temporarily set to 1
Gcv = Kpv*(s+Kiv)/s; %wc = 50 [rad/s], P.M = 90 [deg]

Kpi = 1; Kii = 1;        % temporarily set to 1
Gci = Kpi*(s + Kii)/s; %wc = 4800 [rad/s], P.M = 60 [deg]
```

```
Vref = Vin/D1;
% sisotool Double-loop configuration
T = sisoinit (6);
T.F.Value = Vref;
T.G1.Value = G1;      % current loop G1 = (iL/d)
T.C2.Value = Gci;     % current controller PI = (d/ie)
T.G2.Value = G3;      % voltage loop tf. = (Vc/iL)
T.C1.Value = Gcv;     % voltage controller PI = (Iref/Ve)
sisotool (T);
```

The MATLAB step response of the DC link voltage is depicted in Figure 7. As expected from the design requirements, the waveform of the DC link exhibits no overshoot. The rising time, defined as the time from 10% to 90% of the waveform, is measured as 0.0443 s, while the settling time is found to be 0.0812 s.

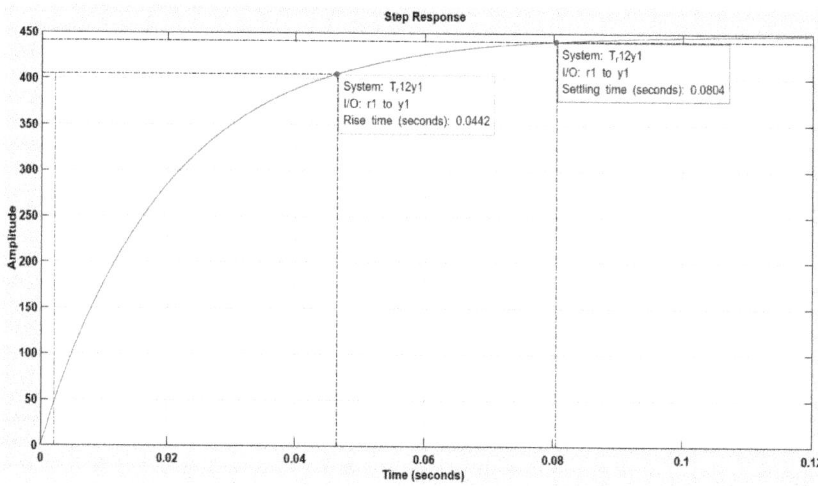

Figure 7. Voltage step response characteristics in MATLAB.

The capacitor balancing controller can be designed from the following equations:

$$i_{C1} = -\frac{v_{c1} + v_{c2}}{R} + i_L(1 - v_{cont1})$$

$$i_{C2} = -\frac{v_{c1} + v_{c2}}{R} + i_L(1 - v_{cont2}) \tag{20}$$

From the above equations, the following relationship can be obtained:

$$i_{C1} - i_{C2} = i_L(v_{cont1} - v_{cont2}) = 2\Delta d i_L \tag{21}$$

The capacitor voltage difference equation over duty difference can then be rewritten as follows:

$$G_{\Delta C} = \frac{\Delta v_c(s)}{\Delta d(s)} = \frac{2}{sC} i_L \tag{22}$$

where, $\Delta v_c(s) = v_{c1} - v_{c2}$.

The closed-loop transfer function of the capacitor voltage difference is given as follows:

$$\frac{\Delta v_c(s)}{\Delta v_{c_{ref}}(s)} = \frac{\frac{2k_p}{C} i_L}{s + \frac{2k_p}{C} i_L} \tag{23}$$

where, $\Delta v_{c_{ref}}(s) = 0$.

The controller configuration of the balancing controller is illustrated in Figure 8a, while its real implementation is depicted in Figure 8b. The controller gain K_p is set to 0.1 based on the design using MATLAB SISOTOOL.

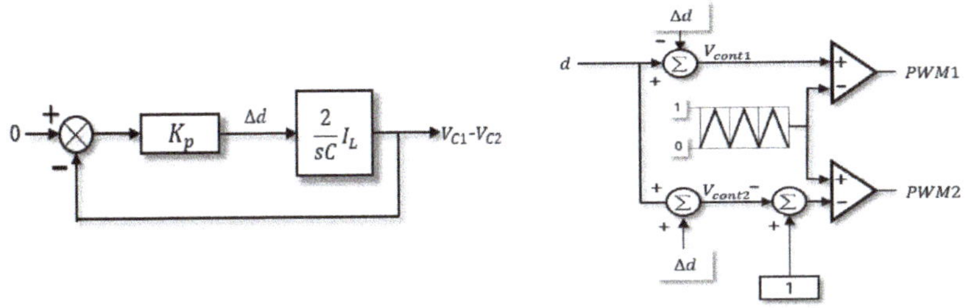

(**a**) controller configuration (**b**) structure of the balancing controller

Figure 8. Capacitor balancing controller.

4. Simulation

The controller was designed using the parameter values shown in Table 2, and the controller design was verified by the PSIM simulation.

Table 2. Controller system parameter.

Parameter	Value
Input Voltage (V_{in})	310 V
Output Voltage (V_{dclink})	450 V
Inductance (I_L)	2.4 mH
Capacitance (C_1, C_2)	2400 µF
Resistance (R)	50 Ω
Switching frequency (f_s)	20 KHz

Figure 9 displays the waveforms of the AC input voltage and AC current from the double-loop controller. It is evident that the power factor (PF) was stably controlled at 0.9932, which is very close to unity.

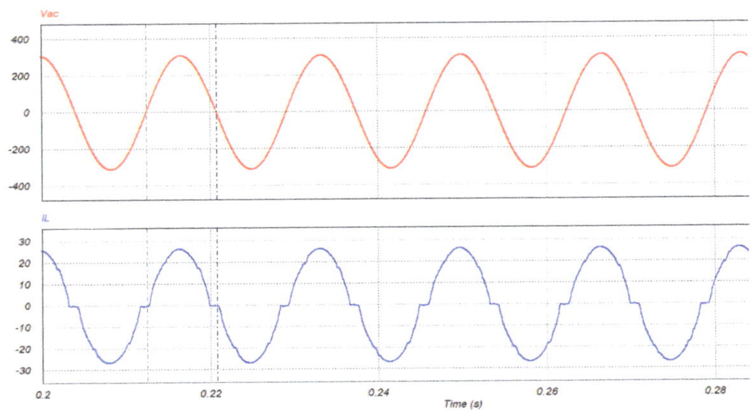

Figure 9. The waveforms of the AC input voltage and AC current.

In Figure 10, the waveforms of the voltage step-up from 450 V to 550 V are presented. Both the output voltage and current are observed to be stably controlled in the boosting section. The actual voltage waveform closely matches the design requirements, except for the 120 Hz ripple components, as shown in Figure 10.

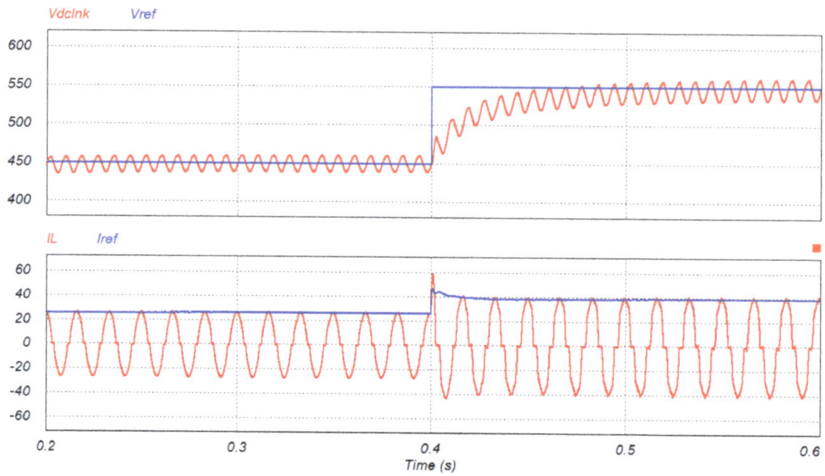

Figure 10. The waveforms of voltage step-up (PSIM simulation).

The control performance for the load variation is shown in Figure 11. This waveform demonstrates the control performance when the load resistance is changed from 25 Ω to 50 Ω and back. Comparing Figures 9 and 10 confirms that the control remains stable even after load changes.

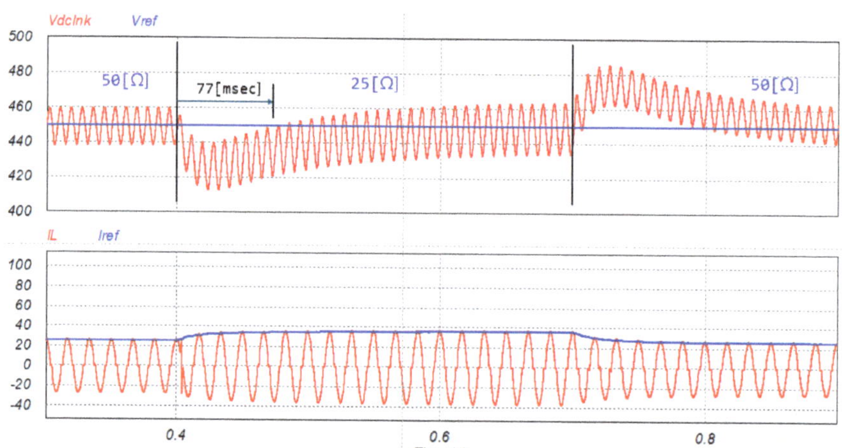

Figure 11. TL PFC converter control performance waveform for load variation.

The waveforms of the balancing controller are depicted in Figure 12. Capacitors C_1 and C_2 are set to 1200 [μF] and 2400 [μF], respectively. The difference in capacitance results in a voltage difference between the capacitors, as shown in Figure 12a. This imbalance can be controlled using the balancing controller, the waveform of which is illustrated in Figure 12b.

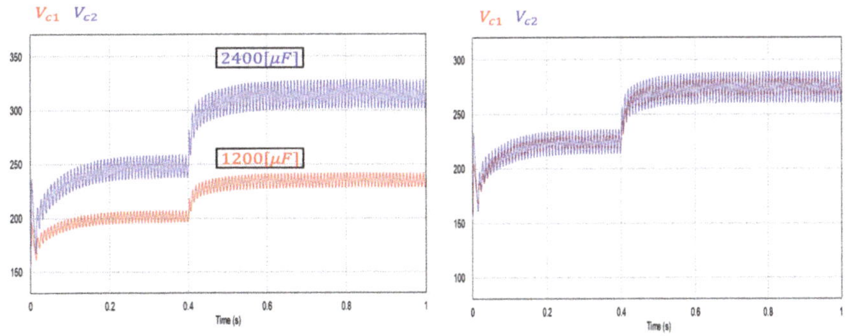

(**a**) Without balancing controller (**b**) with balancing controller

Figure 12. The waveforms of the balancing controller.

5. Experiments

The hardware is constructed based on the simulation environment. The Texas Instruments' 320F28069 DSP is utilized for the micro-controller, and the circuit components used in the experiment are configured to match the values presented in Table 1. The detailed setup of the experiment is depicted in Figure 13. While single-phase AC power at 220 V is used in the PSIM simulation, due to limitations in the experimental system's power, a reduced voltage rating of 110 V is employed during the experiments.

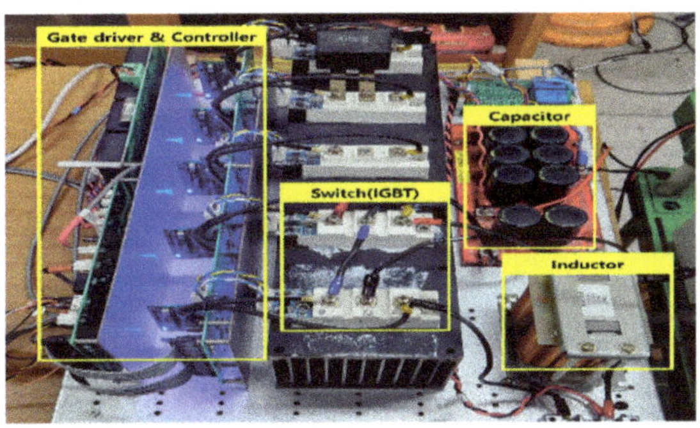

Figure 13. Hardware setup used in the experiment.

First, the step-up and step-down control for output voltage response characteristics are depicted in Figures 14 and 15, respectively. Figure 14 illustrates the result of boosting the output voltage from 150 V to 210 V, with both voltage and current stably controlled without overshoot. Similarly, Figure 15 displays the waveform resulting from stepping down the output voltage from 210 V to 150 V, where both voltage and current are observed to be stably controlled without overshoot.

Figure 16 shows the waveform when changing the load resistance from 50 Ω to 25 Ω, and Figure 17 shows the waveform when changing it from 25 Ω to 50 Ω. In both cases, it was observed that the voltage was stably controlled. However, with regard to the current, fluctuations were initially observed, but the control stabilized within a short period of time.

The waveforms obtained through the actual experiment closely match the PSIM simulation results, confirming the theoretical validity of the proposed modeling and controller design through experimentation.

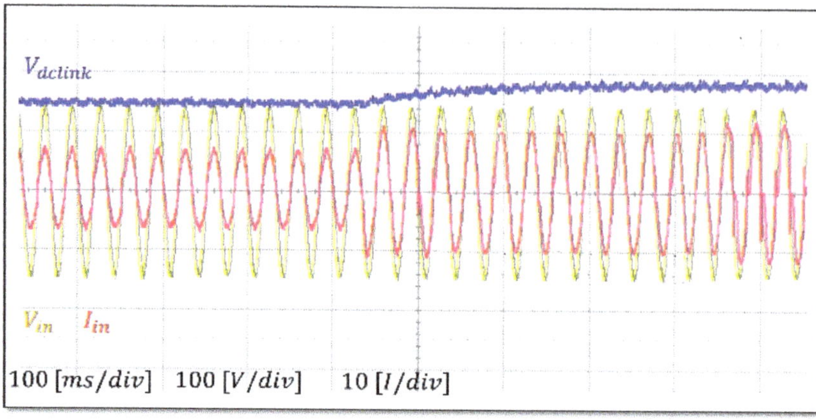

Figure 14. Voltage and current waveforms for reference voltage from 150 V to 210 V.

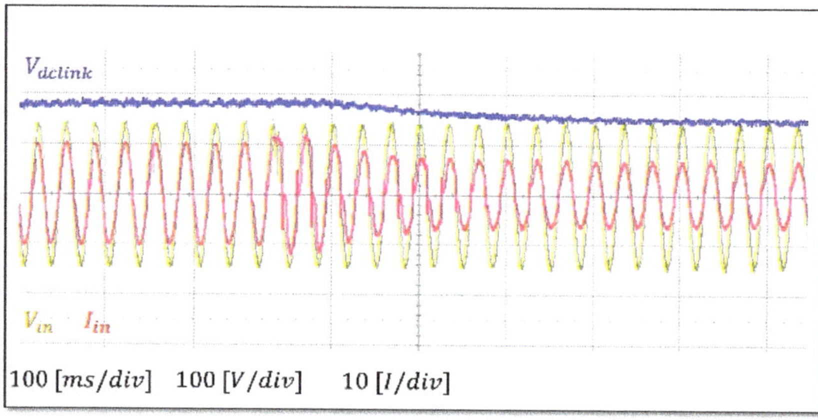

Figure 15. Voltage and current waveforms for reference voltage from 210 V to 150 V.

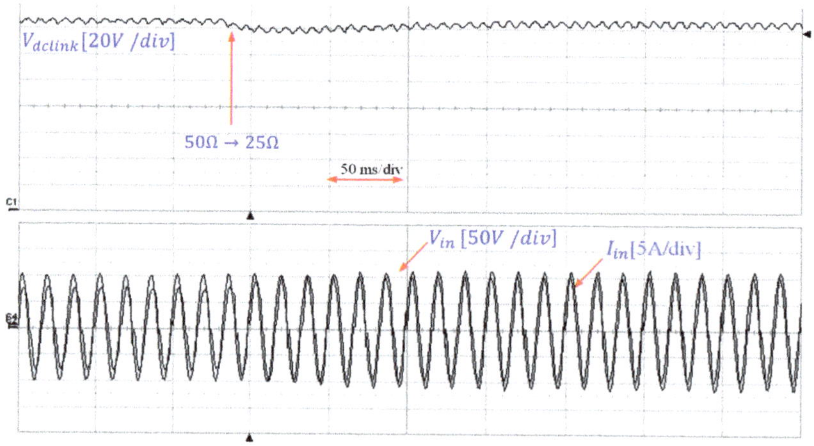

Figure 16. Waveforms for the load change (50 → 25 [Ω]).

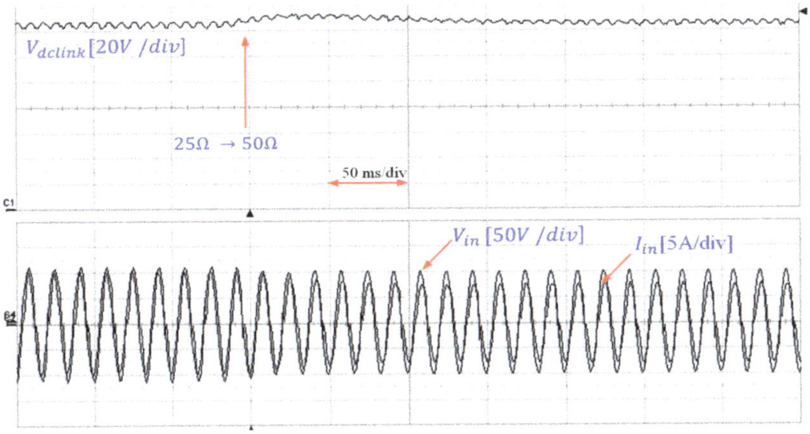

Figure 17. Waveforms for the load change (25 → 50 [Ω]).

6. Conclusions

The paper proposes a double-loop controller design technique to enhance the modeling of a 3-level power factor correction (PFC) converter. The research aims to address challenges encountered in existing single-phase PFC circuits, particularly concerning the presence of a 120 Hz ripple voltage. Prior studies have not adequately addressed double-loop controller design or voltage ripple rejection in this context. Therefore, this paper presents a novel approach by applying a 3-level structure to achieve these goals.

The proposed technique involves configuring the obtained state space expression of the transfer function to be applicable across the full duty range. Subsequently, a double-loop controller is designed using MATLAB SISOTOOL. To evaluate the effectiveness of the proposed controller, verification is conducted through PSIM simulation. Moreover, the proposed controller is implemented in practice, and the efficacy of the system design is confirmed through experimental validation.

The primary contribution of this paper is demonstrating the controllability of the double-loop controller for a three-level PFC converter under varying operating points. This highlights the capability of the proposed controller to maintain operation across the entire duty cycle range, achieving unified control. This achievement distinguishes our work, as previous research has not explored this comprehensive aspect.

Author Contributions: Equally contributed. All authors have read and agreed to the published version of the manuscript.

Funding: This research was supported by "Regional Innovation Strategy (RIS)" through the National Research Foundation of Korea(NRF) funded by the Ministry of Education(MOE)(2021RIS-001). This work was supported by the Human Resources Development of the Korea Insitute of Energy Technology Evaluation and Planning(KETEP) grant funded by the Korea government. (No. 20224000000070).

Data Availability Statement: Data are contained within the article.

Conflicts of Interest: The authors declare no conflict of interest.

References

1. Lee, K.-M.; Kim, I.-S. An Study on the Improved Modeling and Double-loop Controller Design for Three-Level Boost Converter. *Trans. Korean Inst. Power Electron.* **2020**, *25*, 442–450.
2. Pahlevani, M.; Eren, S.; Bakhshai, A.; Jain, P. A DC-bus voltage control technique for grid-connected AC/DC converters used in smart energy storage units. In Proceedings of the 2015 IEEE 6th International Symposium on Power Electronics for Distributed Generation Systems (PEDG), Aachen, Germany, 22–25 June 2015; pp. 1–7.
3. Lee, J.-Y.; Chae, H.-J. 6.6-kW Onboard Charger Design Using DCM PFC Converter With Harmonic Modulation Technique and Two-Stage DC/DC Converter. *IEEE Trans. Ind. Electron.* **2014**, *61*, 1243–1252. [CrossRef]

4. Ryan, R.T.; Hayes, J.G.; Morrison, R.; Hogan, D. Digital control of an interleaved BCM boost PFC converter with fast transient response at low input voltage. In Proceedings of the 2017 IEEE Energy Conversion Congress and Exposition (ECCE), Cincinnati, OH, USA, 1–5 October 2017. [CrossRef]
5. Kim, J.-S.; Kang, J.-I.; Cho, J.-H.; Lee, J.-H.; Ahn, H.-M. DC-link voltage fluctuation compensating algorithm for PFC input voltage magnitude variation. *Power Electron. Conf.* **2023**, 465–466.
6. Ciobotaru, M.; Rossé, A.; Bede, L.; Karanayil, B.; Agelidis, V.G. Adaptive Notch Filter Based Active Damping for Power Converters Using LCL Filters. In Proceedings of the IEEE 7th International Symposium on Power Electronics for Distributed Generation Systems (PEDG'16), Vancouver, BC, Canada, 27–30 June 2016.
7. Yang, D.; Choi, D.; Shin, J.; Choi, B. Stability Analysis of Boost PFC Considering the External Impedance. *Power Electron. Conf.* **2023**, 255–257.
8. Jeong, S.-W.; Lim, J.-H.; Lee, D.-I.; Hyeon, Y.-J.; Lee, S.-Y.; Kim, W.-J.; Youn, H.-S. PFC DC/DC Integrated Converter for High Density of On-Board Charger. *Power Electron. Conf.* **2023**, 550–551.
9. Kim, S.-H.; Cha, H.-N.; Kim, H.-G.; Choi, B.-C. Clamp-mode Three-level High Voltage Gain Boost Converter using Coupled Inductor. *Trans. Korean Inst. Power Electron.* **2012**, *17*, 500–506. [CrossRef]
10. Barazarte, R.Y.; González, G.G. Design of a Two-Level Boost Converter. In Proceedings of the Eleventh LACCEI Latin Acmerican and Caribbean Conference for Engineering and Technology, Cancun, Mexico, 14–16 August 2013; pp. 1–8.
11. Kim, I.S. *Power Conversion Controller Design (Using PSIM and MATLAB)*; Hongneung: Seoul, Republic of Korea, 2023.
12. Keskin, A.; Pal, K.; Hancioglu, E. Resistorless first-order all-pass filter with electronic tuning. *AEU Int. J. Electron. Commun.* **2008**, *62*, 304–306. [CrossRef]
13. Song, M.-G.; Lee, W.-C. A Study on the DC-Link miniaturization and the Reduction of output current distortion rate by Reducing the Effect of 120Hz Ripple Voltage on Photovoltaic Systems. *Power Electron. Conf.* **2020**, 166–167.
14. Sim, W.; Jo, J.; Kim, Y.; Cha, H. Reactive power control of single-phase reactive power compensator for distribution line. *Trans. Korea Inst. Power Electron.* **2020**, *25*, 73–78.
15. Ben Haddi, S.; Zugari, A.; Zakriti, A.; Achraou, S. Design of a Band-Stop Planar Filter for Telecommunications Applications. *Procedia Manuf.* **2020**, *46*, 788–792. [CrossRef]
16. Zhang, R.; Cardinal, M.; Szczesny, P.; Dame, M. A grid simulator with control of single-phase power converters in D-Q rotating frame. In Proceedings of the 2002 IEEE 33rd Annual IEEE Power Electronics Specialists Conference, Cairns, Australia, 23–27 June 2002; pp. 1431–1436.
17. Park, H.C.; Kim, I.S. Optimal Gain Design Method of the 3 Phase Boost Converter. *Trans. Korean Inst. Power Electron.* **2017**, *22*, 1–8. [CrossRef]

electronics MDPI

Article

Robust PI-PD Controller Design: Industrial Simulation Case Studies and a Real-Time Application

Fadi Alyoussef [1,*], Ibrahim Kaya [2] and Ahmad Akrad [3]

1 Institute of Natural and Applied Sciences, Dicle University, 21280 Diyarbakır, Turkey
2 Electrical and Electronics Engineering Department, Dicle University, 21280 Diyarbakır, Turkey; ikaya@dicle.edu.tr
3 ESTACA'Lab, Ecole Supérieure des Techniques Aéronautiques et de Construction Automobile, 53000 Laval, France; ahmad.akrad@estaca.fr
* Correspondence: 16805505@ogr.dicle.edu.tr

Abstract: PI-PD controllers have superior performance compared to traditional PID controllers, especially for controlling unstable and integrating industrial processes with time delays. However, computing the four tuning parameters of this type of controller is not an easy task. Recently, there has been significant interest in determining the tuning rules for PI-PD controllers that utilize the stability region. Currently, most tuning rules for the PI-PD controller are presented graphically, which can be time-consuming and act as a barrier to their industrial application. There is a lack of analytical tuning guidelines in the literature to address this shortfall. However, the existing analytical tuning guidelines do not consider a rigorous design approach. This work proposes new robust analytical tuning criteria based on predefined gain and phase margin bounds, as well as the centroid of the stability region. The proposed method has been tested using various simulation studies related to a DC–DC buck converter, a DC motor, and a heat exchanger. The results indicate that the proposed tuning rules exhibit strong performance against parameter uncertainty with minimal overshoots. Furthermore, the suggested technique for simultaneous control of yaw and pitch angles has been tested in a real-time application using the twin rotor multi-input multi-output system (TRMS). Real-time results indicate that, compared to other methods under investigation, the suggested approach provides nearly minimal overshoots.

Keywords: PI-PD controller; stability region; robust control; centroid point; DC–DC buck converter; DC motor; heat exchanger; twin rotor multi-input multi-output system

1. Introduction

The second order plus time delay (SOPTD) system representation is extensively utilized in the literature because it accurately represents the dynamics of numerous practical industrial applications such as temperature controllers and DC motors. However, long-time delays complicate such systems' control problems as these delays cause uncertainty and slow responses with larger overshoots. Further, the presence of poles on the right-hand side or at the origin of the s-plane adds another complexity to the control design problem. Thus, a robust control approach is required to regulate such systems [1].

PI-PD controllers are strong candidates relative to PID controllers to solve the control mentioned above difficulties [2]. They use the PD part in an inner feedback loop to shift the poles of a process to a more desirable location so that the PI part can control it more effectively in the forward path [3]. Compared to typical PI/PID controllers, PI-PD controllers perform better, especially when it comes to controlling unstable and integrated processes. They also provide the designer more flexibility to enhance closed-loop performance because PI-PD incorporates four adjustable parameters [4]. However, because four controller parameters need to be adjusted, constructing a PI-PD controller is a difficult task [5]. Only a few applications have made use of the PI-PD controller due to the difficulty of calculating those four parameters [6]. In addition, compared to PID controllers, this controller's design has received comparatively less study attention [7]. Therefore, more research on this type of controller is still necessary.

Utilizing the stability region approach to calculate all stabilizing controller gains is a widely researched topic (see [8–11] for examples). Since it provides straightforward and useful tuning guidelines and gets around the challenge of tuning the PI-PD controller, many researchers have recently explored using the centroid point of the stability region, which can be calculated by using the centroid of convex stability region (CCSR) [12] or the weighted geometrical center (WGC) approaches [13,14], to adjust the parameters of PI-PD controllers. Additionally, it has been suggested that the controller evaluated using the centroid point may produce robustness against parameter fluctuations, faster perturbation rejection, and improved set-point tracking [15,16]. WGC and CCSR have just lately been used to graphically adjust the PID controller's gains, which are intended to control systems with non-integer order delay [17]. Additionally, a comparison study of the PI-PD controller tuning techniques employing CCSR and WGC has recently been conducted [14]. Because CCSR and WGC approaches have only been applied to specified transfer functions in the above-mentioned literature, they share the common drawback of requiring the design process to be repeated whenever the transfer function changes. This might be time-consuming for controller engineers and might require mathematical knowledge. The user-selected step size to be employed in the design phase causes a computational load, which is a unique drawback of the WGC approach [12]. Additionally, if the time lag is particularly significant, the WGC approach may yield an incorrect centroid location because of its sensitivity to the chosen step size [18]. The graphical way of determining the centroid location is a drawback unique to the CCSR approach [4]. Additionally, novel methods for determining a fractional order controller's stability region's centroid point have recently been introduced [19,20]. In addition, the centroid point has recently been used in the controller parameter design for DC–DC boost converters [21]. Nevertheless, the centroid point is still obtained graphically in all of the aforementioned stated approaches, which is laborious.

To remove the above-mentioned disadvantages of centroid point approaches, analytical tuning rules have been recently suggested [3,4,16,18,22]. However, these approaches have not considered incorporating a centroid point based on a robust design method. In fact, since the original work that formulated the stability region's formulas for time delay systems and PI-PD controllers, as documented in Ref. [23], the majority of researchers have tuned these controllers using graphical approaches without taking analytical ones into account. Furthermore, robust design techniques as previously discussed have not taken into account analytical methodologies. Therefore, the main goal of this paper is to design a robust tuning rule based on the centroid point, which can specifically withstand the system parameter uncertainties and provide acceptable settling times and overshoots. The primary contributions made by this work are:

1. Relative to the centroid point approaches reported in Refs. [3,4,16,18,22], the proposed method incorporates a robust design approach based on predefined gain and phase margin boundaries, which gives the designer more flexibility to obtain the desired control performances.
2. Relative to the centroid point approaches reported in Refs. [12,14,19,20], the proposed methods are analytical and not graphical, which saves time and is easy to implement on the industrial level.
3. Compared to the centroid point derived based on a robust design technique for DC–DC converter control published in Ref. [24], the suggested approach is analytical and does not call for any graphical adjustments. Furthermore, any system that can be represented as a stable SOPTD can use the suggested approach.

To conclude, proposing an analytical robust design method for designing the gains of the PI-PD controller is the main contribution of this paper. This is the layout for the rest of the paper. The next part explains the PI-PD controller structure. The suggested methodology is expounded upon in Section 3. Section 4 describes the simulation and the real-time results. The last section deals with conclusions.

2. PI-PD Controller Structure

Every action taken by the PID regulator in a traditional closed-loop system occurs via the forward path. This can cause an unwanted occurrence known as a derivative kick. To solve this issue, the PD part of PI-PD regulators is moved to an internal feedback loop, which moves the poles of the plant transfer function to a more favorable location where the PI component operating in the onward path might more adequately control it [4,18]. The PI-PD regulator's architecture is shown in Figure 1. The inner loop is composed of the transfer functions of the plant, $G(s)$, and the PD controller, $C_{PD}(s)$, whereas the external loop is composed of the transfer functions of the inner loop and the PI regulator, $C_{PI}(s)$. The transfer functions $C_{PD}(s)$ and $C_{PI}(s)$ of the PI-PD regulator are written as follows:

$$C_{PD}(s) = k_f + k_d s \quad (1)$$

$$C_{PI}(s) = k_p + \frac{k_i}{s} \quad (2)$$

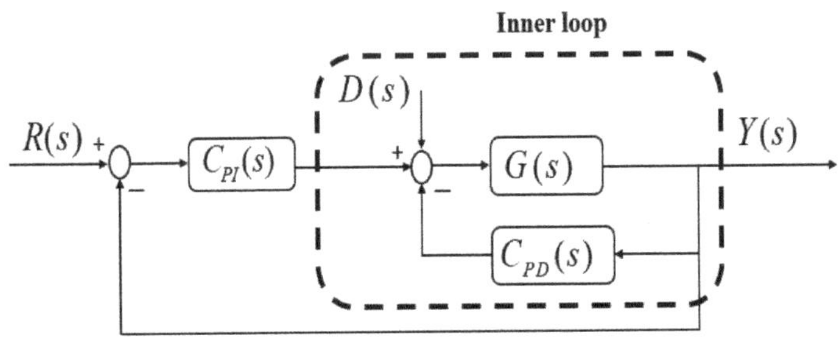

Figure 1. PI-PD controller.

The behavior of many industrial plants can be described as a second-order transfer function with time delays as follows:

$$G(s) = \frac{Ke^{-\tau s}}{as^2 + bs + c} \quad (3)$$

In the above equation, the time delay, the process gain, and the coefficients of the transfer function are τ, K, a, b, and c, respectively.

3. Proposed Approach

The gain-phase margin tester, $T(s) = Ae^{-\phi s}$, is added to the forward path to find boundaries in the general stability boundary corresponding to the predetermined gain and phase margins. Figure 2 shows the structure of the PI-PD controller after adding the tester [23].

The following closed-loop characteristic equation can be used to characterize the inner feedback loop shown in Figure 2, which is made up of the PD controller transfer function, $C_{PD}(s)$, and the process transfer functions, $G(s)$:

$$\Delta(s) = 1 + C_{PD}(s)G(s) \Rightarrow \Delta(s) = k_f Ke^{-\tau s} + k_d Kse^{-\tau s} + as^2 + bs + c \quad (4)$$

In the parameter space methodology, the root of a stable polynomial has three ways of crossing over the imaginary axis and becoming unstable. The real root boundary, the infinite root boundary, and the complex root boundary are the three boundaries that these ways define [4]. To find the real root boundary, we solve Equation (4) with $s = 0$ and $\Delta(s) = 0$. Therefore, $k_f = -c/K$ defines the boundary. Since the controller parameters are absent from the coefficient of s^2, the infinite root boundary cannot be part of the stability region. To find the complex root boundary, we enter $s = j\omega$, $e^{-j\tau\omega} = \cos(\tau\omega) - j\sin(\tau\omega)$, and $j = \sqrt{-1}$ into Equation (4) as given below:

$$\Delta(s) = -a\omega^2 + jb\omega + c + k_f K\cos(\tau\omega) - jk_f K\sin(\tau\omega) + jk_d K\omega\cos(\tau\omega) + k_d K\omega\sin(\tau\omega) = 0 \quad (5)$$

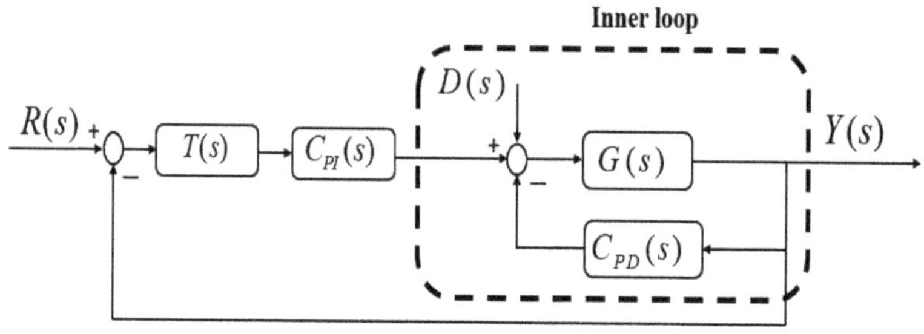

Figure 2. PI-PD controller plus gain-phase margin tester.

The imaginary and real components in Equation (5) are then split and equaled to zero. Consequently, the following equations can be obtained:

$$k_d(\omega\cos(\tau\omega)) - k_f(\sin(\tau\omega)) = -\frac{b}{K}\omega \quad (6)$$

$$k_d(\omega\sin(\tau\omega)) + k_f(\cos(\tau\omega)) = \frac{a\omega^2 - c}{K} \quad (7)$$

The expressions that follow can be obtained by solving Equations (6) and (7) for k_d and k_f:

$$k_d = \frac{1}{K}\left(-b\cos(\tau\omega) + (a\omega - \frac{c}{\omega})\sin(\tau\omega)\right) \quad (8)$$

$$k_f = \frac{1}{K}\left((a\omega^2 - c)\cos(\tau\omega) + b\omega\sin(\tau\omega)\right) \quad (9)$$

The inner feedback loop's stability region is defined by Equations (8) and (9) and the line $k_f = -c/K$. By varying ω throughout the range of $[\varepsilon \quad \omega_{PD}]$, Equations (8) and (9) are utilized to construct the stability region. ε is a small number that is used to avoid dividing by zero in Equation (8), while ω_{PD} is found by taking the first root of the equation: $(a\omega^2 - c)\cos(\tau\omega) + b\omega\sin(\tau\omega) = -c$. The adjusting point of the PD controller can be found using the mathematical relations in the range of $[e \quad f]$ [18]:

$$\overline{x} = \frac{1}{f-e}\int_e^f z(x)dx \quad (10)$$

$$\overline{y} = \frac{1}{2(f-e)}\int_e^f g(x)dx \quad (11)$$

The integrations provided in Equations (10) and (11) should be used to analytically calculate the settings of the PD controller. Also, the same mathematical relations will be used later for finding the settings of the PI controller. Considering $f = \omega_{PD}$, $e = 0$, and $z(x) = k_d$ which is given in Equation (8), and $g(x) = k_f$ which is given in Equation (9), the gains of the PD controller are analytically computed using Equations (10) and (11). Thus, the following expressions are obtained and used to find the setting of this controller:

$$\overset{*}{k_d} = \frac{1}{K\omega_{PD}}\left(\frac{a(\sin(\tau\omega_{PD}) - \tau\omega_{PD}\cos(\tau\omega_{PD}))}{\tau^2} - \frac{b\sin(\tau\omega_{PD})}{\tau} - c\left(\frac{\tau^6\omega_{PD}^5}{600} - \frac{\tau^4\omega_{PD}^3}{18} + \tau^2\omega_{PD}\right)\right) \quad (12)$$

$$\overset{*}{k}_f = \frac{1}{2K\omega_{PD}} \left(\begin{array}{c} \frac{b(\sin(\tau\omega_{PD}) - \tau\omega_{PD}\cos(\tau\omega_{PD}))}{\tau^2} + \\ \frac{(\tau^2(a\omega_{PD}^2 - c) - 2a)\sin(\tau\omega_{PD}) + 2a\tau\omega_{PD}\cos(\tau\omega_{PD})}{\tau^3} \end{array} \right) \tag{13}$$

Here, it is important to illustrate that the first three terms of the Taylor series are used for approximation $\frac{\tau \sin(\tau\omega)}{\tau\omega}$ and consequently achieving the integration action in Equation (13). The first three terms of the Taylor series are selected as they can provide good computation accuracy without introducing mathematical complexity. To compute the controller parameters using Equations (12) and (13), one should first identify the parameters of the SOPTD system reported in Equation (3). After that, ω_{PD} should be computed by taking the first root of the equation: $(a\omega^2 - c)\cos(\tau\omega) + b\omega\sin(\tau\omega) = -c$. This can be achieved by using a scientific package such as MATLAB.

To find the PI controller parameters $\overset{*}{k}_i$ and $\overset{*}{k}_p$, the outer loop's characteristic equation is given below:

$$\Delta(s) = \left(\begin{array}{l} 1 + G_{PD}(s)G(s) + Ae^{-j\phi}G_{PI}(s)G(s) = \\ as^3 + bs^2 + cs + AKk_pse^{-j(\tau\omega+\phi)} + AKk_ie^{-j(\tau\omega+\phi)} + K\overset{*}{k}_fse^{-\tau s} + K\overset{*}{k}_ds^2e^{-\tau s} \end{array} \right) \tag{14}$$

Afterward, the real root boundary, the infinite root boundary, and the complex root boundary are determined. For the outer loop, the real root boundary is found by solving Equation (14) for $s = 0$ and $\Delta(s) = 0$. Thus, this boundary is defined by $k_i = 0$. The infinite root boundary cannot be a component of the stability zone since the controller's settings are missing from the coefficient of s^3. The complex root boundary is found by entering $s = j\omega$, $e^{-j\tau\omega} = \cos(\tau\omega) - j\sin(\tau\omega)$, $e^{-j\phi} = \cos(\phi) - j\sin(\phi)$, and $j = \sqrt{-1}$ into Equation (14). After that, imaginary and real components are split to obtain the following equations:

$$k_p(\omega\sin(\tau\omega + \phi)) + k_i(\cos(\tau\omega + \phi)) = \frac{b}{AK}\omega^2 - \frac{\overset{*}{k}_f}{A}\omega\sin(\tau\omega) + \frac{\overset{*}{k}_d}{A}\omega^2\cos(\tau\omega) \tag{15}$$

$$k_p(\omega\cos(\tau\omega + \phi)) - k_i(\sin(\tau\omega + \phi)) = \frac{a}{AK}\omega^3 - \frac{c}{AK}\omega - \frac{\overset{*}{k}_f}{A}\omega\cos(\tau\omega) - \frac{\overset{*}{k}_d}{A}\omega^2\sin(\tau\omega) \tag{16}$$

The complex root boundary's equations are given by solving Equations (15) and (16) for k_p and k_i as follows:

$$k_p = \left(\begin{array}{l} \frac{b}{AK}\omega^2\cos(\tau\omega + \phi) + \frac{b}{AK}\omega\sin(\tau\omega + \phi) - \frac{\overset{*}{k}_f}{A} + \sin(\tau\omega)\sin(\tau\omega + \phi) \\ -\frac{\overset{*}{k}_f}{A}\cos(\tau\omega)\cos(\tau\omega + \phi) + \frac{\overset{*}{k}_d}{A}\omega\cos(\tau\omega)\sin(\tau\omega + \phi) - \frac{\overset{*}{k}_d}{A}\omega\sin(\tau\omega)\cos(\tau\omega + \phi) \\ -\frac{c}{AK}\cos(\tau\omega + \phi) \end{array} \right) \tag{17}$$

$$k_i = \left(\begin{array}{l} -\frac{a}{AK}\omega^3\sin(\tau\omega + \phi) + \frac{c}{AK}\omega\sin(\tau\omega + \phi) + \frac{b}{AK}\omega^2\cos(\tau\omega + \phi) \\ +\frac{\overset{*}{k}_f}{A}\omega\cos(\tau\omega)\sin(\tau\omega + \phi) - \frac{\overset{*}{k}_f}{A}\omega\sin(\tau\omega)\cos(\tau\omega + \phi) + \frac{\overset{*}{k}_d}{A}\omega^2\sin(\tau\omega)\sin(\tau\omega + \phi) \\ +\frac{\overset{*}{k}_d}{A}\omega^2\cos(\tau\omega)\cos(\tau\omega + \phi) \end{array} \right) \tag{18}$$

The line $k_i = 0$ and Equations (17) and (18) are used to determine the stability region of the outer loop. To plot Equations (17) and (18), ω is changed throughout the range of $\left[0 \quad \omega_{PI}\right]$, where ω_{PI} is found by equating Equation (18) to zero and taking the first solution. After applying the formulae from Equations (10) and (11) to Equations (17) and (18), correspondingly, the settings of the PI controller are determined as follows:

$$\overset{*}{k}_p = \frac{1}{\omega_{PI}} \left(\begin{array}{c} \frac{\left(2a\tau^2\omega_{PI}^2 - 2c\tau^2 + 2b\tau - 4a\right)\sin(\tau\omega_{PI} + \phi) + \left(4a\tau - 2b\tau^2\right)\omega_{PI}\cos(\tau\omega_{PI} + \phi)}{2AK\tau^3} \\ + \frac{Kk_d\tau^3\omega_{PI}^2\sin(\phi) - 2Kk_f\tau^3\omega_{PI}\cos(\phi)}{2AK\tau^3} + \frac{\left(2c\tau^2 - 2b\tau + 4a\right)\sin(\phi)}{2AK\tau^3} \end{array} \right) \tag{19}$$

$$\overset{*}{k_i} = \frac{1}{2\omega_{PI}} \left(\begin{array}{l} \frac{\left(\left(6b\tau^3 - 18a\tau^2\right)\omega_{PI}^2 + 6c\tau^2 - 12b\tau + 36a\right)\sin(\tau\omega_{PI} + \phi)}{6AK\tau^4} \\ + \frac{\left(6a\tau^3\omega_{PI}^3 + \left(-6c\omega_{PI}\tau^3 + 12b\tau^2 - 36a\tau\right)\omega_{PI}\right)\cos(\tau\omega_{PI} + \phi)}{6AK\tau^4} \\ + \frac{+2Kk_d\tau^4\cos(\phi)\omega_{PI}^3 + 3Kk_f\tau^4\sin(\phi)\omega_{PI}^2}{6AK\tau^4} - \frac{\left(6c\tau^2 - 12b\tau + 36a\right)\sin(\phi)}{6AK\tau^4} \end{array} \right) \tag{20}$$

One can find the controller parameters using the flowchart shown in Figure 3. Here, it is important to mention that no approximations have been used to find Equations (19) and (20).

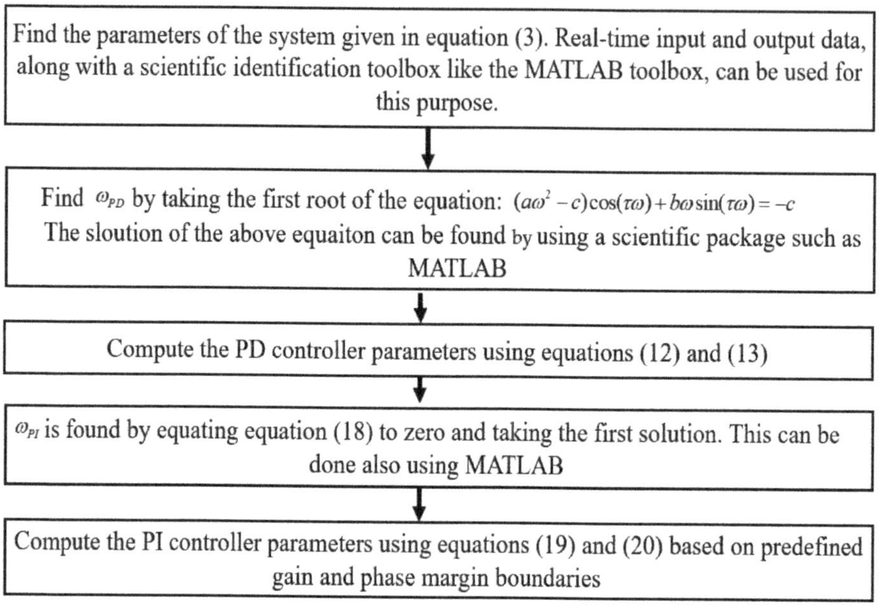

Figure 3. The flowchart for finding the PI-PD parameters using the proposed method.

4. Simulation and Real-Time Case Studies

Three simulation case studies and a real-time application will be shown in this section. The PD controller's derivative component with the form $C_{PD}(s) = k_f + k_d s/(t_f s + 1)$, where $t_f = (k_d)/(10k_f)$, is used for noise rejection and practical application. To ensure a fair comparison with the methods in the literature, the derivative filter will also be introduced to all literature techniques under consideration. The examined approaches are numerically evaluated using the gain margin, the phase margin, the integral of absolute error (IAE), $IAE = \int_0^\infty |e|dt$, and the total variation (TV), $TV = \int_0^\infty (u(t+1) - u(t))dt$.

Case study 1: A DC–DC buck converter, which is shown in Figure 4, is studied here. The identified transfer function of this system is $G(s) = \frac{8.87e^{2\times 10^{-4}\ s}}{7.0341\ \times 10^{-7}s^2 + 0.0017s + 1}$ [25]. The suggested method for different predefined gain and phase margin boundaries is applied to find out the controller settings using the procedures shown in Figure 3. Table 1 summarizes these settings' values as well as the PID controller's values reported in Reference [25].

Figures 5 and 6 show the stability region with computed centroid points using the proposed method for inner and outer loops, respectively. Figure 6 shows the stability region for three predefined scenarios of the gain and phase margin boundaries, which are $(A = 1 \quad \phi = 0^\circ)$, $(A = 1.5 \quad \phi = 0^\circ)$, and $(A = 1.6 \quad \phi = \pi/4^\circ)$. The scenario corresponding to $(A = 1 \quad \phi = 0^\circ)$ means that no gain and phase margin boundaries are included in the design of the centroid point. First, the described controllers are assessed using the system parameters' nominal values. Specifically, at t = 0 s, the system is given a unity step input. Additionally, at t = 0.02 s, a disturbance with an amplitude of 0.3 is introduced into the closed-loop system. Figure 7 illustrates the output voltage responses of the assumed scenarios along with the PID [25] and their control signals. From the figure, all the proposed scenarios give better results in terms of generating small overshoots relative to the PID controller [25], which appears to consume more energy with possible oscillation, as is clear in Figure 7b. In fact, the overshoot and observed oscillations of the PID controller [25] might be a source of instability in practical implementations. On the other hand, the proposed method does not have an overshoot, which might make it more suitable for real-time applications. Additionally, Figure 7a shows that the proposed scenarios $(A = 1 \quad \phi = 0^{\circ})$ relative to other scenarios and PID [25] quickly rejects the injected disturbances into the closed loop with an amplitude of 0.3 at t = 0.02 s. Thus, one might conclude that the proposed method, due to the freedom in the selection of the predefined values of the gain and phase margins, has the flexibility to be tuned by control engineers to generate the desired performances according to the application in hand. This can be seen numerically in Table 1, in which one of the proposed scenarios has at least the smallest values for IAE or TV. The optimal values of the gain and phase margins lie in the ranges $\begin{bmatrix} 2 & 6 \end{bmatrix}$ and $\begin{bmatrix} 30^{\circ} & 60^{\circ} \end{bmatrix}$, respectively [4]. Therefore, it seems from Table 1 that the proposed method with all the investigated scenarios provides optimum values of the gain and phase margin relative to the PID controller [25], which has a gain margin of less than 2. The numerical observations, given in Table 1, illustrate again how the proposed method is flexible in terms of providing different desired control performances.

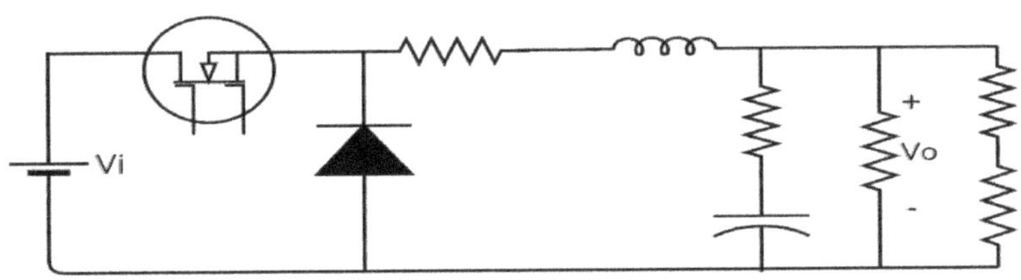

Figure 4. DC–DC buck converter circuit [25].

Table 1. The controller parameters and performance measures for all the studied methods for case study 1.

Method	Controller Parameters				Real Phase and Gain Margins		IAE		TV	
	k_f	k_d	k_p	k_i	Gain Margin	Phase Margin	Servo	Reg.	Servo	Reg.
Proposed $(A = 1 \quad \phi = 0^{\circ})$	0.539	0.0002	0.104	605.619	2.69	43.48°	0.0011	0.00051	0.00000061	0.0000011
Proposed $(A = 1.5 \quad \phi = 0^{\circ})$	0.539	0.0002	0.069	395.578	2.74	49.23°	0.0016	0.00074	0.00000036	0.0000010
Proposed $(A = 1.6 \quad \phi = \pi/4^{\circ})$	0.539	0.0002	0.170	235.089	2.58	45.18°	0.0028	0.0013	0.00000071	0.0000012
PID [25]	-	0.0004	0.545	626.437	1.75	49.93°	0.0013	0.00066	0.00000041	0.0000022

To evaluate the effectiveness of the approaches under evaluation in the event of a situation where the system parameters change, another test is carried out. This test is particularly important to show how the suggested method gives a robust performance against parameter uncertainties. To show the benefits of incorporating the predefined gain and phase margin boundaries into the proposed method, the system parameters τ, a, b, and c are changed by 500%. The system's parameters are changed by 500% since the reported controllers start to show a noticeable influence. Figure 8 displays the findings of this experiment. It is evident that, in contrast to PID and scenario $(A = 1 \quad \phi = 0^{\circ})$, which display oscillatory behavior and the potential to lose stability, scenarios $(A = 1.5 \quad \phi = 0^{\circ})$

and $(A = 1.6 \quad \phi = \pi/4^{\circ})$ maintain system stability with a very small overshoot. To sum up, the suggested scenarios examined in Figures 7 and 8 demonstrate that the suggested approach gives the control engineer the flexibility and opportunity to choose the most suitable control performance, potentially weighing fast-tracking against robust performance in the face of system parameter fluctuations. This possibility is not available in the centroid tuning methods reported in Refs. [3,4,16,18,22], which forms the main contribution of this paper in addition to the analytical nature of the proposed method.

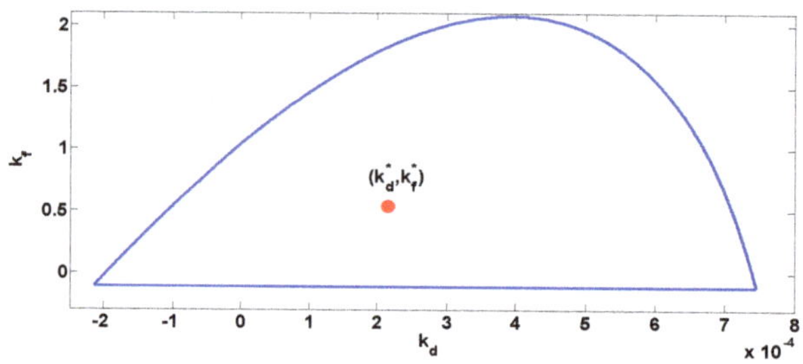

Figure 5. Stability region of the PD controller for Case study 1 for the nominal parameters.

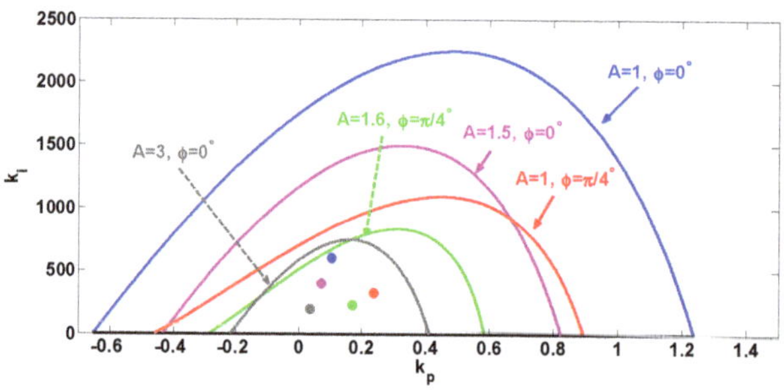

Figure 6. Stability region of the PI controller for Case study 1 for the nominal parameters.

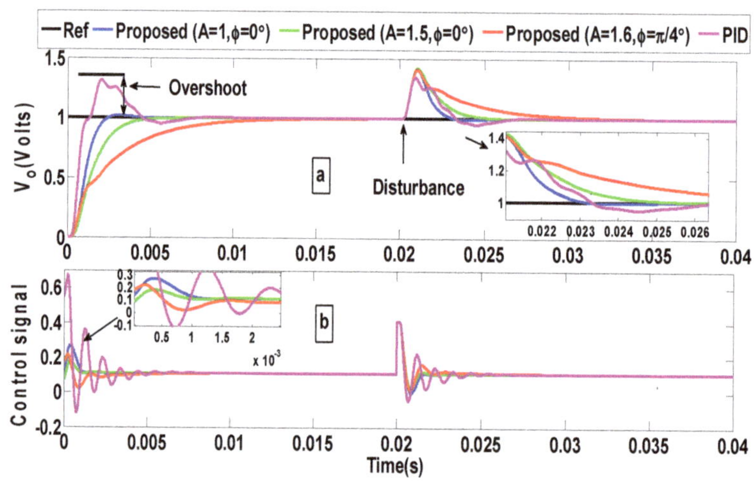

Figure 7. (**a**) Closed-loop responses for DC–DC buck converter under a 0.3 disturbance amplitude and for the nominal parameters; (**b**) corresponding control signals.

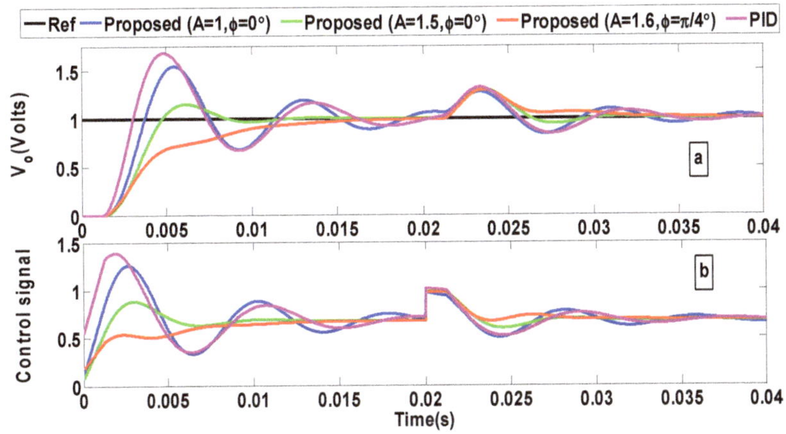

Figure 8. (**a**) Closed-loop responses for DC–DC buck converter under a 0.3 disturbance amplitude and 500% model uncertainty; (**b**) corresponding control signals.

Case study 2: A position control problem for a DC motor is considered in this case study. The identified transfer function of the DC motor position is $G(s) = \frac{0.9843e^{-0.02s}}{0.0651s^2+s}$ [26]. The parameters of the proposed method for the assumed values of $(A = 1.3 \quad \phi = 0^\circ)$ are computed using the steps given in Figure 3. The fractional order PI controller (FOPI) [26] and the PI-PD controller adjusted by CCSR [4] are used to compare the performance with the suggested approach. Table 2 contains a list of all the reported controller settings values together with the values of performance measures.

Table 2. The controller parameters and performance measures for all the studied methods for case study 2.

Method	Controller Parameters					Real Phase and Gain Margins		Step Input				Variable Input	
								IAE		TV		IAE	TV
	k_f	k_d	k_p	k_i	μ	Gain Margin	Phase Margin	Servo	Reg.	Servo	Reg.	Servo	Servo
Proposed (A = 1.3 $\phi = 0^\circ$)	38.838	1.866	8.175	300.739	-	2.52	37.3°	0.13	0.003	0.005	0.0005	0.13	0.0048
PI-PD [4]	47.715	2.703	29.818	823.207	-	1.67	22.9°	0.10	0.002	0.058	0.0017	0.09	0.058
FOPI [26]	-	-	12.103	123.761	1.6	0.63	35.9°	0.66	0.115	0.005	0.0007	0.66	0.0048

First of all, the investigated methods are tested using the nominal parameters and by applying a unity response and unity disturbance at t = 0 s and t = 3 s, respectively. Figure 9 displays the outcomes for this test. It is clear from the figure that FOPI [26] has poor performance with oscillatory response and long settling times. On the other hand, the proposed method offers a quick response with zero overshoot and a similar disturbance rejection capability relative to PI-PD [4], which has an oscillatory transient response. Also, the PI-PD [4] consumes more energy relative to the proposed controller as seen in the control signals given in Figure 9b.

The second test for the methods under investigation includes introducing parameter perturbations. By introducing 30% uncertainties in all system parameters, the benefits of incorporating the predetermined gain and phase margin boundaries into the design procedures of the proposed method are demonstrated in Figure 10. From the figure, the suggested strategy is the only one that can continue to function well even when there are fluctuations in the system parameters, while other approaches lose their stability entirely. This makes the proposed method suitable for controlling systems that have changing parameters.

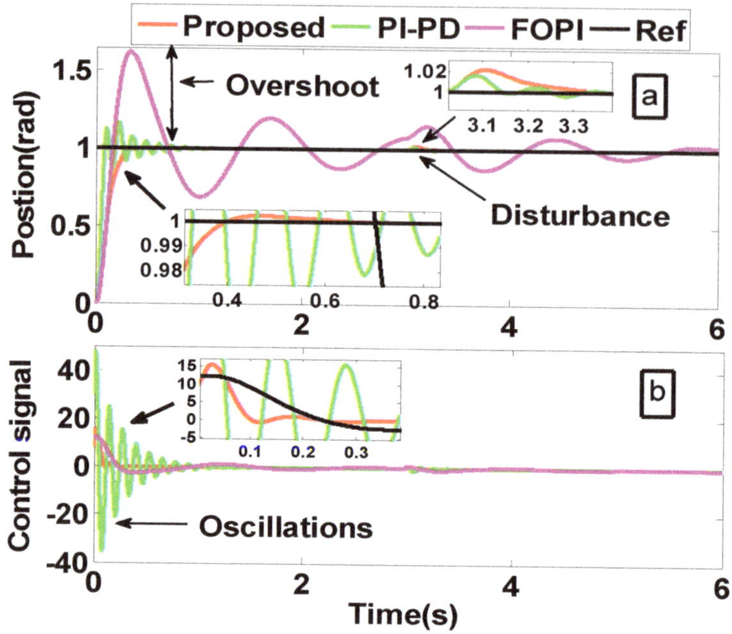

Figure 9. (**a**) Closed-loop responses for DC motor for a step input and under a unity disturbance amplitude for the nominal parameters; (**b**) corresponding control signals.

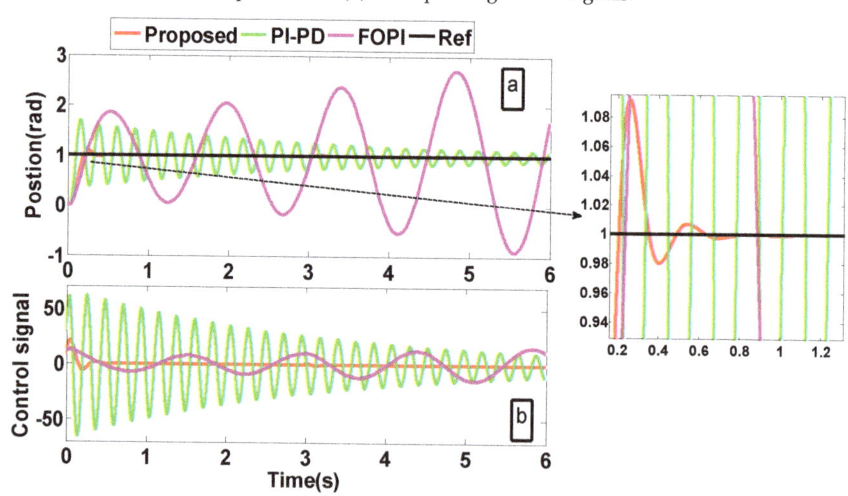

Figure 10. (**a**) Closed-loop responses for DC motor for a step input and under a unity disturbance amplitude and 30% model uncertainty; (**b**) corresponding control signals.

The third test for the reported methods is performed using a variable input consisting of square and sinusoidal signals, as shown in Figure 11a. One can see from the figure that the proposed method has decent performance without introducing oscillations or overshoots. As can be observed from the IAE readings for the variable input given in Table 2, the PI-PD controller [4] provides the fewest tracking errors; but, as Figure 11a,b makes evident, it exhibits oscillatory transient responses. This might lead to instability in real-time scenarios. Also, the FOPI controller [26] has a response with overshoots and oscillations, as noticeable in Figure 11a. Thus, one can conclude that only the proposed method can give suitable performance without overshoots and oscillations and with reasonable tracking of errors.

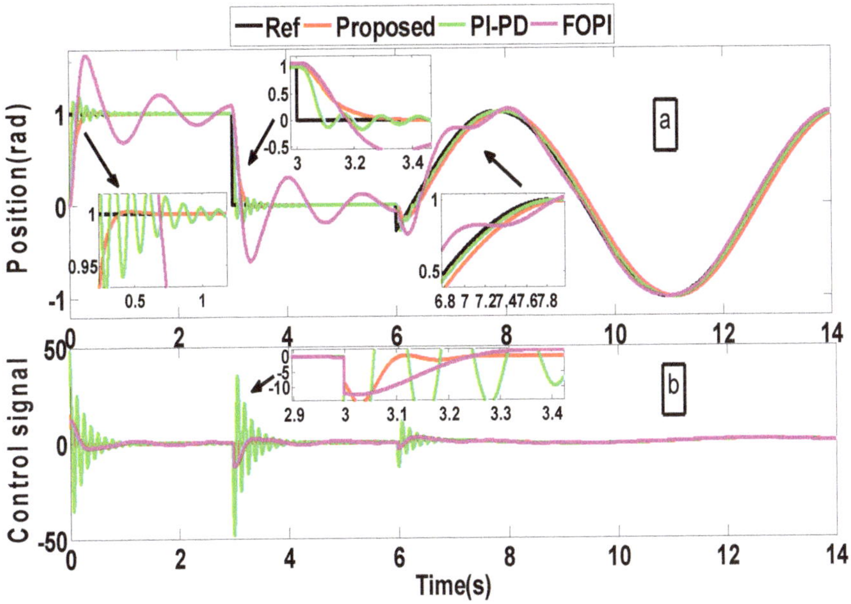

Figure 11. (**a**) Closed-loop responses for DC motor for a variable input and under the nominal parameters; (**b**) corresponding control signals.

The TV values, listed in Table 2, show that the proposed method has the smoothest control signal for servo or regulatory responses for both types of control signals. The IAE values presented in the same table demonstrate that, for either servo or regulatory responses, the suggested approach offers a trade-off performance between the strategies described in Refs. [4,26] for all the input types. However, different gain and phase margins might be chosen if control engineers want a different level of intended performance. This is because the suggested method has the flexibility to do so as shown in case study 1. Also, the proposed method has optimal values of the gain and phase margins relative to the other methods as clear from Table 2. To conclude, the above observations demonstrate, yet again, the flexibility, robustness, and importance of the suggested approach.

Case study 3: An industrial temperature control based on a heat exchanger, displayed in Figure 12, is considered under this case study. A shell-and-tube system is used to transfer the heat from a primary to a secondary flow. The process control goal is to maintain the secondary flow point's temperature by regulating the primary flow. The system's recognized transfer function is $G(s) = \frac{0.002e^{-3s}}{s}$ [2]. By following the methods shown in Figure 3, one can calculate the settings of the suggested technique for assumed scenarios of $(A = 1 \quad \phi = 0°)$ and $(A = 1.3 \quad \phi = 0°)$. The values of these settings are summarized in Table 3, together with the reported values of the PI-PD controller plus the Smith predictor reported in reference [2]. PI-PD-SP is the abbreviation for the PI-PD controller plus the Smith predictor [2] in this case study. The goal of the comparison between the proposed method and PI-PD-SP is to show that the proposed method can perform well in the presence of long time delays, even though it has not incorporated another technique for handling the long time delay such as the Smith predictor.

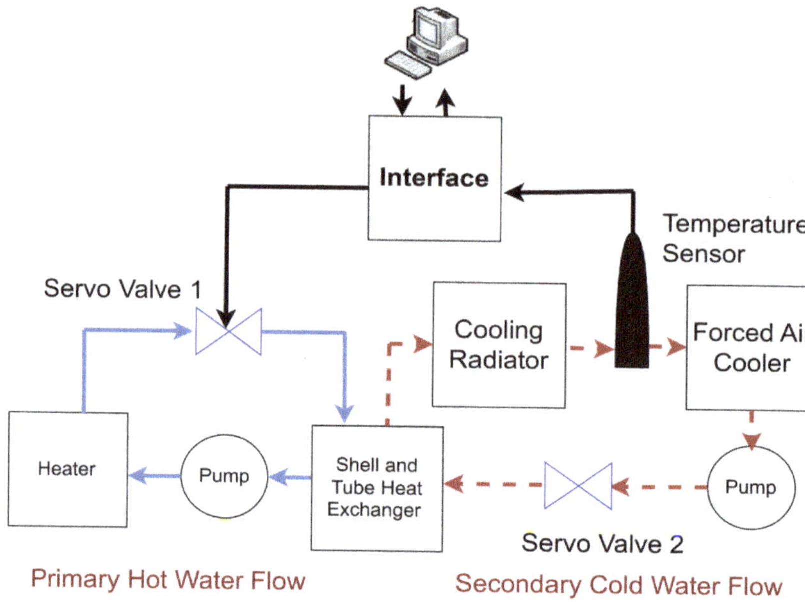

Figure 12. Diagrammatic representation of temperature control using a heat exchanger.

Table 3. The controller parameters and performance measures for Case study 3.

Method	Controller Parameters				Real Phase and Gain Margins		Step Input				Square Input	
							IAE		TV		IAE	TV
	k_f	k_d	k_p	k_i	Gain Margin	Phase Margin	Servo	Reg.	Servo	Reg.	Servo	Servo
Proposed (A = 1 $\phi = 0°$)	83.34	0	22.762	8.113	2.18	32.63°	10.272	0.370	0.014	0.0013	30.817	0.028
Proposed (A = 1.5 $\phi = 0°$)	83.34	0	15.175	5.409	2.45	39.87°	15.408	0.555	0.009	0.0011	46.203	0.042
PI-PD-SP [2]	130.15	188.326	26.011	25.601	1.93	27.85°	5.184	0.120	0.029	0.0018	15.553	0.086

Different tests are performed to test the proposed method. In the first test, a unity step input and disturbance with an amplitude of 3 are applied at t = 0 s and t = 80 s, respectively. The outcomes of this test are shown in Figure 13. The second test includes introducing variable input in the form of square input as shown in Figure 14. The third test is similar to the first one but after introducing 70% model uncertainty in the time delay. Its results are shown in Figure 15. The following observations are noted based on Figures 13–15, as well as the performance metrics listed in Table 3:

1. As seen in Figures 13 and 14, the suggested scenario, $(A = 1 \quad \phi = 0°)$, provides somewhat similar responses and results to the PI-PD-SP controller [2], even though it does have an additional mechanism to tackle long time delays.
2. In contrast to PI-PD-SP [2], which completely loses stability under 70% variations in the time delay as clear in Figure 15, the suggested scenarios $(A = 1 \quad \phi = 0°)$ and $(A = 1.3 \quad \phi = 0°)$ can keep the system stable with a satisfactory performance. Also,

the robustness of the proposed method can be numerically seen from the real values of the gain and phase margin provided in Table 3.

3. As can be observed from the TV values listed in Table 3, the suggested scenarios provide smoother control signals. Furthermore, as shown in Figures 13b and 14b, the suggested method uses less control power for the transient response than the PI-PD-SP [2].

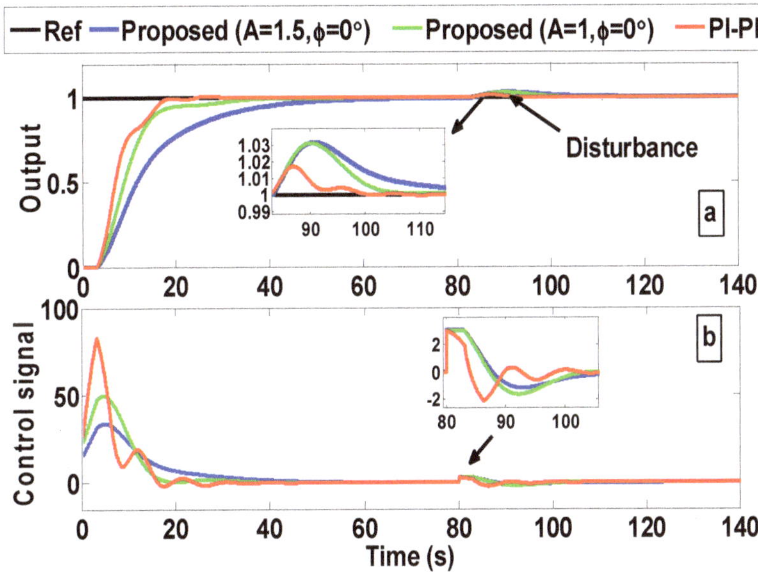

Figure 13. (**a**) Closed-loop responses for the temperature control for a step input and under the nominal parameters; (**b**) corresponding control signals.

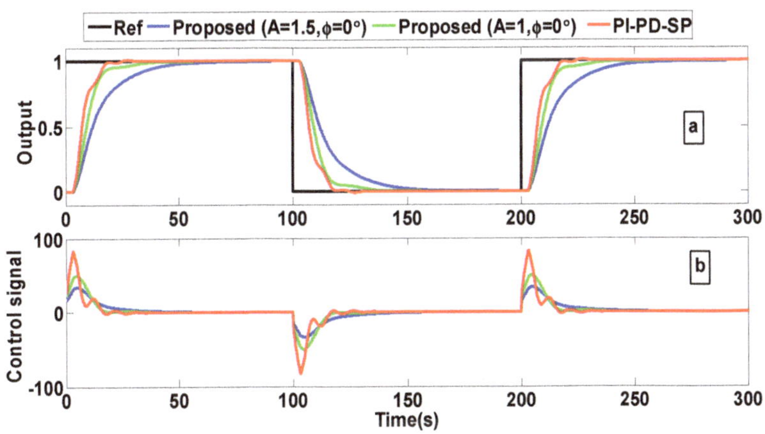

Figure 14. (**a**) Closed-loop responses for the temperature control for a square input and under the nominal parameters; (**b**) corresponding control signals.

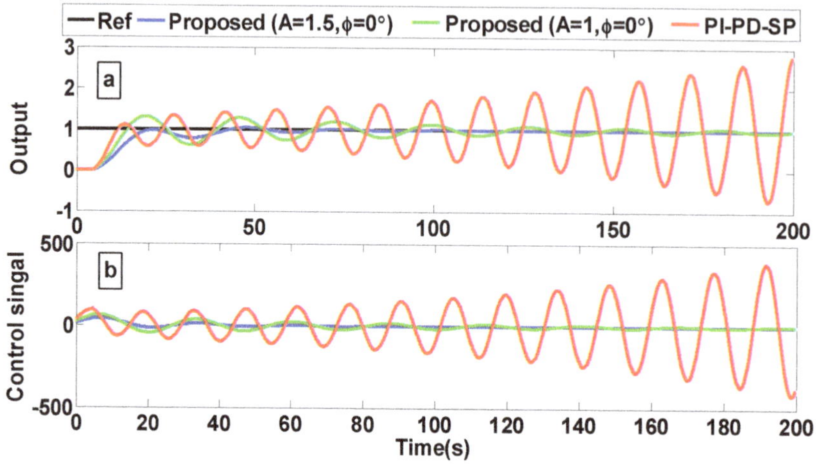

Figure 15. (**a**) Closed-loop responses for the temperature control under a unity disturbance amplitude and 70% time delay model uncertainty; (**b**) corresponding control signals.

Experimental results: Here, the proposed technique is tested using a real-time trial on a TRMS. The dynamics of the nonlinear TRMS system are comparable to those of a helicopter. As seen in Figure 16, it comprises two DC motors that drive the pitch and yaw propellers at the ends of a beam. The yaw rotor produces horizontal revolution on the yaw axis, whilst the pitch rotor is in charge of the upward motion on the pitch axis. There is a noticeable cross-coupling between the rotor's actions, with each rotor affecting both angles. The yaw transfer function of yaw angle is identified as $G(s) = \frac{0.10e^{-0.3227s}}{0.95965s^2+s}$ [4]. On the other hand, the transfer function of the pitch angle is identified using the input and output data as $G(s) = \frac{0.32e^{-s}}{2.347s^2+0.1s+1}$.

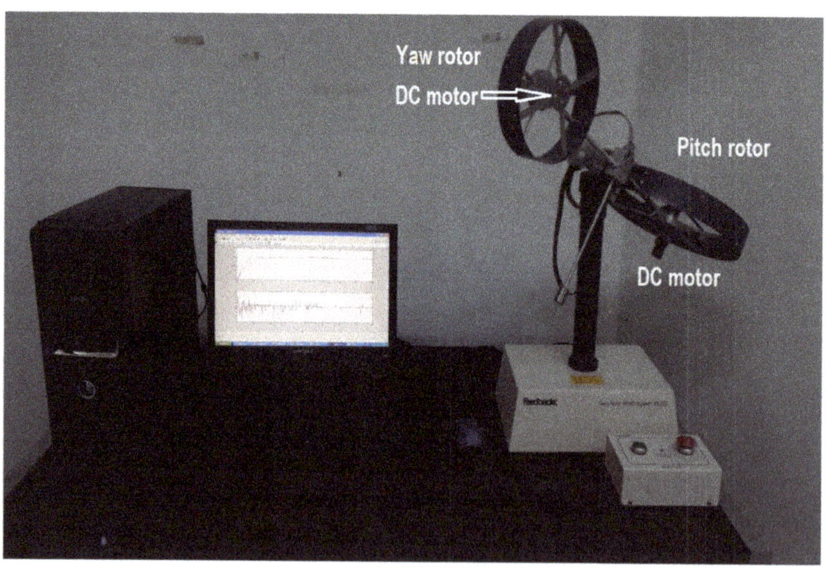

Figure 16. TRMS setup.

The suggested method in Figure 3 is used to compute the settings of the yaw and pitch PI-PD controllers using the assumed scenarios $(A = 1 \quad \phi = 0°)$ and $(A = 1 \quad \phi = \pi/3.7°)$, respectively. The proposed controller is implemented using the diagram shown in Figure 17. The PID controller described in Ref. [27] is contrasted with the suggested approach. Table 4 displays the IAE values together with the parameters for every method that has been reported. To ensure a fair comparison, it is vital to note that every controller evaluated is subjected to identical implementation conditions. An example of these conditions might include the absence of external disturbances such as wind.

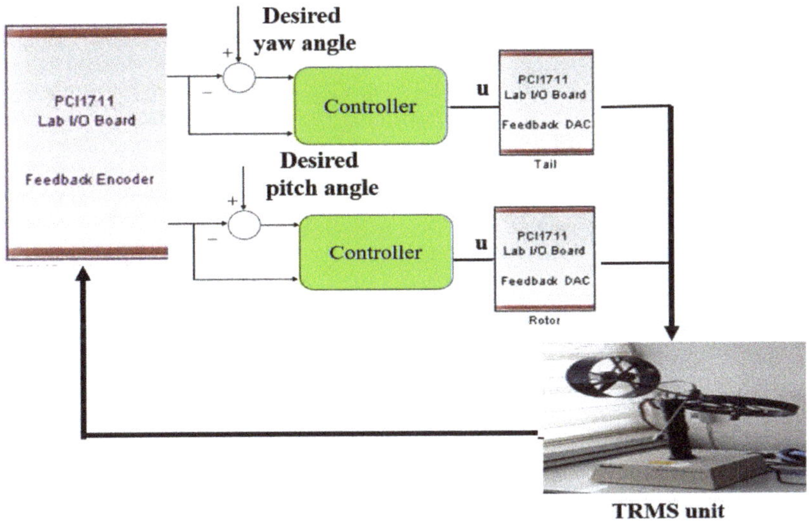

Figure 17. Diagram for implementing the PI-PD controller.

Table 4. The controller parameters and IAEs for all the studied methods for the real-time test.

	Method	Controller Parameters				IAE
		k_f	k_d	k_p	k_i	
Proposed	Yaw controller $(A = 1 \quad \phi = 0°)$	22.4563	16.6002	6.1352	14.0555	0.9711
	Pitch controller $(A = 1 \quad \phi = \pi/3.7°)$	0.1122	2.6569	0.1959	1.1341	1.7484
PID [27]	Yaw controller	-	5	2	1	1.7774
	Pitch controller	-	10	5	6	1.2817

The real-time outcomes for concurrently controlling the pitch and yaw angles are shown in Figures 18 and 19. The simulation outcomes for the suggested method are also displayed in Figures 18a and 19a. Interestingly, the proposed method yields somewhat similar real-time and simulation results, even though the TRMS is a highly nonlinear system with coupling effects that may be challenging to model using low-order transfer functions. The suggested approach exhibits rapid responses with shorter settling times and overshoot, particularly for the yaw response, as seen in Figure 18. Additionally, Figures 18b and 19b demonstrate how much less energy the suggested controller uses than the PID controller [27]. Furthermore, even though the suggested approach is primarily suggested for single-input single-output systems, the suggested controller could continue to function well considering the existence of coupling influences between yaw and pitch angles. This is explained by the flexibility of the suggested approach, which enables the designer to use predetermined gain and phase margin boundaries to achieve the desired performance.

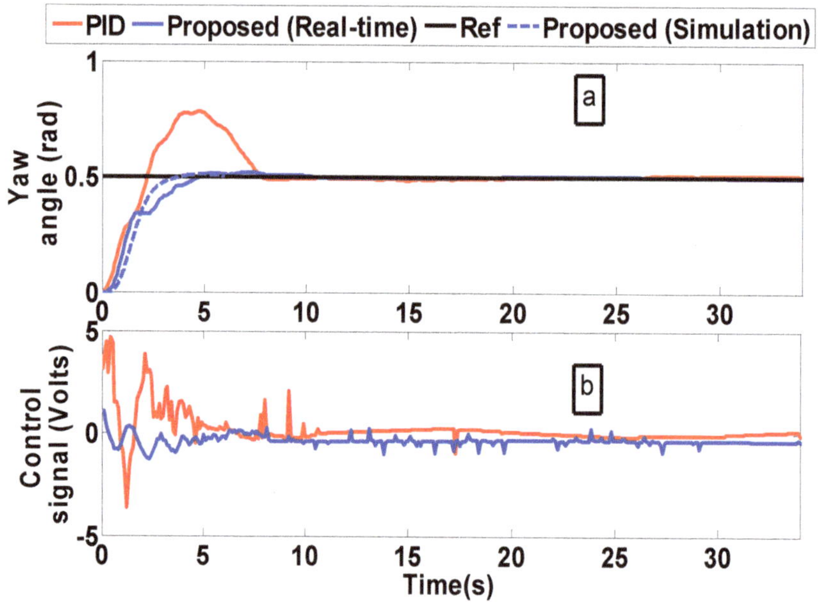

Figure 18. (**a**) Real-time closed-loop yaw responses for TRMS and (**b**) corresponding control signals.

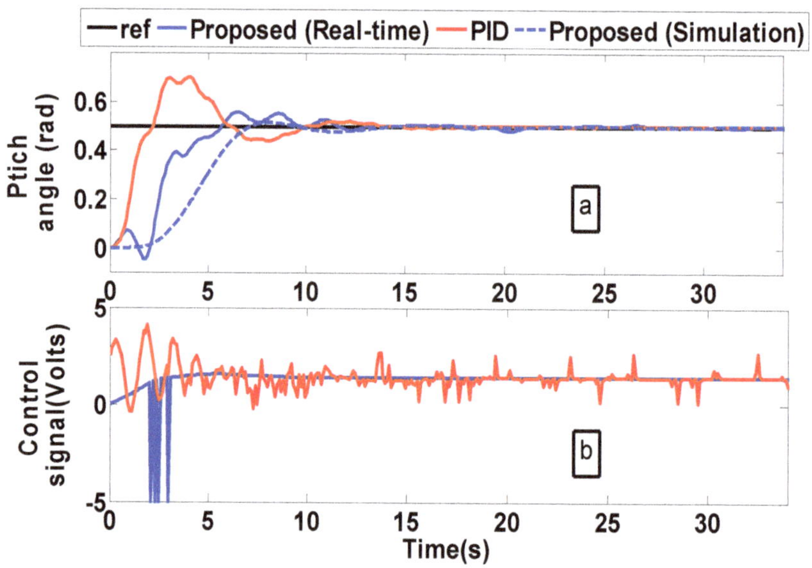

Figure 19. (**a**) Real-time closed-loop pitch responses for TRMS and (**b**) corresponding control signals.

5. Conclusions

A robust centroid point design technique based on predefined phase and gain margins is presented in this paper. In comparison to the existing literature, the suggested approach is both analytical and robust. The control engineer has the flexibility to choose the required control performance according to the suggested approach. The analytical centroid point techniques described in the literature really lack this feature. The results of the simulations demonstrated that the suggested approach produces responses with reduced overshoots and settling times and performs well against parameter uncertainties. These findings are further supported by the results of the real-time test on TRMS, which shows that the proposed methodology can handle the coupling effects to larger levels when utilized for controlling multiple input and multiple output systems even if it is originally only designed to deal with single-input and single-output systems. Nonetheless, the coupling effects continue to influence the pitch angle control performance, albeit to a lesser extent. Future studies will therefore involve extending the suggested approach to handle the coupling effects in multiple-input and multiple-output systems.

Author Contributions: Methodology, F.A. and I.K.; software, F.A.; validation, F.A.; formal analysis, F.A.; investigation, F.A.; resources, F.A.; data curation, F.A.; writing—original draft, F.A.; writing—review and editing, I.K. and A.A.; visualization, F.A.; supervision, I.K; project administration, I.K.; funding acquisition, A.A. All authors have read and agreed to the published version of the manuscript.

Funding: This research received no external funding.

Data Availability Statement: Data is contained within the article.

Acknowledgments: This study was implemented at Dicle University under the research project number: Mühendislik.22.002.

Conflicts of Interest: The authors declare no conflicts of interest.

List of Symbols

Symbol	Explanation
K	Process gain
τ	Time delay
a, b, c	Coefficients of the process transfer function
$G(s)$	Process transfer function
$G_{PD}(s)$	PD controller's transfer function
$G_{PI}(s)$	PI controller's transfer function
k_f	The proportional gain of the PD controller
k_d	The derivative gain of the PD controller
k_p	The proportional gain of the PI controller
k_i	The integral gain of the PI controller
A	Gain margin
θ	Phase margin
ω_{PD}	The critical frequency of the inner loop
ω_{PI}	The critical frequency of the outer loop
$[e \quad f]$	The minimum and maximum integral limits
$\bar{x}$	The centroid coordinate on the x-axis
$\bar{y}$	The centroid coordinate on the y-axis
μ	The degree of integral operator

References

1. Alyoussef, F.; Kaya, I. Improved adaptive dynamic non-singular terminal sliding mode controller with fractional disturbance observer. *Inf. Sci.* **2023**, *641*, 119110. [CrossRef]
2. Singha, P.; Das, D.; Chakraborty, S.; Lloyds Raja, G. Experimentally validated predictive PI-PD control strategy for delay-dominant chemical processes. *Chem. Eng. Sci.* **2024**, *295*, 120197. [CrossRef]
3. Alyoussef, F.; Kaya, I. Novel Tuning Rules for Adjusting the Parameters of the PI-PD Controller for Controlling Unstable Processes. *Adv. Transdiscipl. Eng.* **2024**, *46*, 254–261.
4. Alyoussef, F.; Kaya, I. Simple PI-PD tuning rules based on the centroid of the stability region for controlling unstable and integrating processes. *ISA Trans.* **2023**, *134*, 238–255. [CrossRef] [PubMed]
5. Kaya, I. I-PD Controller Design for Integrating Time Delay Processes Based on Optimum Analytical Formulas. *IFAC-PapersOnLine* **2018**, *51*, 575–580. [CrossRef]
6. Zou, H.; Li, H. Improved PI-PD control design using predictive functional optimization for temperature model of a fluidized catalytic cracking unit. *ISA Trans.* **2017**, *67*, 215–221. [CrossRef]

7. Lloyds Raja, G.; Ali, A. New PI-PD Controller Design Strategy for Industrial Unstable and Integrating Processes with Dead Time and Inverse Response. *J. Control Autom. Electr. Syst.* **2021**, *32*, 266–280. [CrossRef]
8. Mahapatro, S.R.; Subudhi, B. A New H∞ Weighted Sensitive Function-Based Robust Multi-Loop PID Controller for a Multi-Variable System. *IEEE Trans. Circuits Syst. II Express Briefs* **2024**, *71*, 1256–1260. [CrossRef]
9. Jain, C.; Vaishnav, N.; Jain, A.K. MTPA Control of Double Inverter Fed Wound Rotor Induction Motor Drive. *IEEE Trans. Energy Convers.* **2023**, *38*, 2326–2343. [CrossRef]
10. Vaishnav, N.; Jain, A. Stabilizing Sets of Current PI Controllers for IM Drives with and without LC Filter. *IEEE Trans. Energy Convers.* **2022**, *37*, 958–969. [CrossRef]
11. Mehta, U.; Aryan, P.; Raja, G.L. Tri-Parametric Fractional-Order Controller Design for Integrating Systems with Time Delay. *IEEE Trans. Circuits Syst. II Express Briefs* **2023**, *70*, 4166–4170. [CrossRef]
12. Onat, C. A new design method for PI–PD control of unstable processes with dead time. *ISA Trans.* **2019**, *84*, 69–81. [CrossRef] [PubMed]
13. Ozyetkin, M.M.; Onat, C.; Tan, N. PI-PD controller design for time delay systems via the weighted geometrical center method. *Asian J. Control* **2020**, *22*, 1811–1826. [CrossRef]
14. İrgan, H.; Menak, R.; Tan, N. A comparative study on PI-PD controller design using stability region centroid methods for unstable, integrating and resonant systems with time delay. *Meas. Control* **2024**. [CrossRef]
15. Imran Kalim, M.; Ali, A. New tuning strategy for series cascade control structure. *IFAC-PapersOnLine* **2020**, *53*, 195–200. [CrossRef]
16. Alyoussef, F.; Kaya, I. Tuning proportional-integral controllers based on new analytical methods for finding centroid of stability locus for stable/unstable first-order plus dead-time processes. *Proc. Inst. Mech. Eng. Part I J. Syst. Control Eng.* **2022**, *236*, 818–831. [CrossRef]
17. Ozyetkin, M.M. An approximation method and PID controller tuning for systems having integer order and non-integer order delay. *Alex. Eng. J.* **2022**, *61*, 11365–11375. [CrossRef]
18. Alyoussef, F.; Kaya, I. Proportional–integral and proportional–derivative controller design based on analytically computed centroid point for controlling integrating processes. *Proc. Inst. Mech. Eng. Part I J. Syst. Control Eng.* **2023**, *237*, 1045–1065. [CrossRef]
19. Yildirim, B.; Gheisarnejad, M.; Khooban, M.H. A Robust Non-Integer Controller Design for Load Frequency Control in Modern Marine Power Grids. *IEEE Trans. Emerg. Top. Comput. Intell.* **2021**, *6*, 852–866. [CrossRef]
20. Yildirim, B.; Gheisarnejad, M.; Khooban, M.H. A New Parameter Tuning Technique for Noninteger Controllers in Low-Inertia Modern Power Grids. *IEEE J. Emerg. Sel. Top. Ind. Electron.* **2021**, *3*, 279–288. [CrossRef]
21. Kalim, M.I.; Ali, A. A graphical approach for controller design with desired stability margins for a DC–DC boost converter. *IET Power Electron.* **2021**, *14*, 1323–1335. [CrossRef]
22. Çökmez, E.; Kaya, İ. An analytical solution of fractional order PI controller design for stable/unstable/integrating processes with time delay. *Turk. J. Electr. Eng. Comput. Sci.* **2023**, *31*, 626–645. [CrossRef]
23. Tan, N. Computation of stabilizing PI-PD controllers. *Int. J. Control Autom. Syst.* **2009**, *7*, 175–184. [CrossRef]
24. Kumar, S.R.; Anwar, M.N.; Verma, P.; Ram, M.K.; Somanshu, S.R. H∞ Criterion Based PI Controller Design for DC-DC Step-Up Converters. *IEEE J. Emerg. Sel. Top. Ind. Electron.* **2023**, *5*, 402–413. [CrossRef]
25. Ramana, K.V.; Majhi, S.; Gogoi, A.K. Modeling and Estimation of DC-DC Buck Converter Dynamics Using Relay Feedback Output with Performance Evaluation. *IEEE Trans. Circuits Syst. II Express Briefs* **2019**, *66*, 427–431.
26. Lino, P.; Maione, G.; Stasi, S.; Padula, F.; Visioli, A. Synthesis of fractional-order PI controllers and fractional-order filters for industrial electrical drives. *IEEE/CAA J. Autom. Sin.* **2017**, *4*, 58–69. [CrossRef]
27. Alyoussef, F.; Kaya, I. TRMS Experimental Results of New Nonlinear PID Tuned by de Algorithm. In Proceedings of the 2019 3rd International Conference on Applied Automation and Industrial Diagnostics, ICAAID 2019, Elazig, Turkey, 25–27 September 2019.

electronics

MDPI

Article

Advanced Distributed Control of Parallel Resonant CLLC DAB Converters

David Carmona Vicente [1,*], Alba Muñoz Carrero [1], Eduardo Galván Díez [1,*], Juan Manuel Carrasco Solís [1] and Francisco Rodríguez Rubio [2]

[1] Department of Electronic Engineering, University of Seville, 41092 Seville, Spain; amunoz1@us.es (A.M.C.); jmcarrasco@us.es (J.M.C.S.)
[2] Department of Systems Engineering and Automatics, University of Seville, 41092 Seville, Spain; rubio@us.es
* Correspondence: dcarmona1@us.es (D.C.V.); egalvan@us.es (E.G.D.)

Abstract: The integration of hybrid alternating current (AC) and direct current (DC) networks has gained relevance due to the growing demand for more flexible, efficient, and reliable electrical systems. A key aspect of this integration is the parallelization of power converters, which presents several technical challenges, such as current sharing imbalances, circulating currents, and control complexity. This paper proposes a distributed control architecture for parallel resonant CLLC dual active bridge (DAB) converters to address these issues in hybrid AC–DC networks and microgrids. The approach includes a master voltage controller to regulate the output voltage and distributed local current controllers to ensure load balance. The approach minimizes the difference between the output and input voltages, allowing for independent control of power flow. Simulation and experimental results show significant improvements. The system stability has been demonstrated experimentally. Transient response has been improved with response time 80% lower using the feed-forward term. The system maintained stability with current sharing deviations below 3% under full and low load conditions. Finally, scalability is ensured by the proposed distributed controller because the central power controller is not affected by the number of units in parallel used in the application. This solution is suitable for advanced hybrid networks and microgrid applications.

Keywords: resonant converters; dual active bridge (DAB); distributed control; parallel converters; circulating currents; hybrid AC–DC networks; microgrids; solid-state transformers (SSTs); smart grids

1. Introduction

In recent years, microgrids (MGs) have become increasingly important for improving the efficiency, reliability, and flexibility of electrical power systems. MGs integrate distributed generators, energy storage systems, loads, and control units that can operate independently or connected to the main grid, providing increased stability and resilience in the power supply [1]. The evolution of MGs has led to the development of various topologies, including AC, DC, and hybrid grids, which require advanced control strategies to optimally manage distributed resources [2].

Traditionally, secondary and tertiary control layers in MGs have been implemented using centralized control topologies, where a central controller manages all information and makes operational decisions for the microgrid [3]. However, this centralized approach presents several significant limitations, such as vulnerability to single-point failures and a lack of scalability when integrating new resources or adapting to changes in demand [4]. Furthermore, dependence on a single central unit for decision-making can compromise the robustness and responsiveness of the system to unexpected events [5].

Distributed control has emerged as a promising solution to overcome these limitations. Unlike centralized control, distributed control spreads control responsibility between several autonomous agents, which work cooperatively to achieve global objectives [6]. This approach offers several key advantages, such as increased robustness to failures, since the failure of a single agent does not lead to total system breakdown [7]. Moreover, distributed control improves scalability, allowing the seamless integration of new distributed energy resources (DERs), storage systems, and loads without affecting the overall operation of the microgrid [8].

The integration of hybrid AC-DC electrical grids has gained prominence in recent years due to the growing demand for more flexible, efficient, and reliable electrical systems. A key aspect of this integration is the parallelization of power converters, which poses several technical challenges, such as current sharing imbalances, circulating currents, and the complexity of control [9,10]. These issues can significantly affect the performance and reliability of hybrid networks, leading to inefficiencies and operational risks [11].

Various strategies have been proposed to address these challenges, including advanced control schemes, voltage droop control, and current droop control, as well as real-time communication between converters [12]. Recent trends emphasize the use of digital controllers, predictive algorithms, and artificial intelligence to improve system performance and scalability [13]. However, solutions based on centralized controllers, such as programmable logic controllers (PLCs), suffer from slow transient response and communication delays, which limit overall system stability [14].

The CLLC resonant converter has emerged as a promising solution to parallelize dual active bridge (DAB) converters in hybrid AC–DC networks, particularly in high-power applications [15]. The CLLC converter is known for its ability to operate with high efficiency in bidirectional modes, making it ideal for use in solid-state transformers (SSTs). However, parallel operation of multiple CLLC converters introduces significant technical challenges, such as unequal current sharing between converters at both the input and output stages and the precise control of circulating currents [16].

This paper proposes a novel distributed control strategy that eliminates the need for a global controller and instead relies on a local master controller installed on a single converter to regulate the output voltage. The remaining converters are controlled by distributed current controllers that share the load current equally, and feed-forward control is implemented to improve the system response time during transient events.

The remainder of this paper is organized as follows. Section 2 describes the materials and methods used, addressing the modeling of CLLC DAB resonant converters and the system architecture, and presents the traditional control strategies and the proposed distributed control methodology, including the master voltage controller and the distributed current controllers together with the feed-forward term in control. Section 3 shows the experimental results obtained in test configurations, including positive and negative AC power flow, response to nominal power and voltage variations, and the effect of the feed-forward term in the system. Section 4 discusses the impact of the distributed control strategy compared to traditional centralized systems, highlighting the improvements in terms of response time, stability, and current distribution under various load conditions. Finally, Section 5 concludes the paper by summarizing the main contributions and suggesting directions for future research, such as the integration of artificial intelligence to improve system adaptability.

2. Materials and Methods

2.1. Paper Preparation

During the preparation of this work, GenAI tools have been used to improve the paper. CHATGPT V4 has been used for preparing the reference list during the state of the art compilation and to improve English writing. The authors have reviewed and edited all the results and take full responsibility for the content of the publication.

2.2. Modeling of Resonant CLLC DAB Converters

Figure 1 illustrates the topology of a resonant CLLC DAB converter.

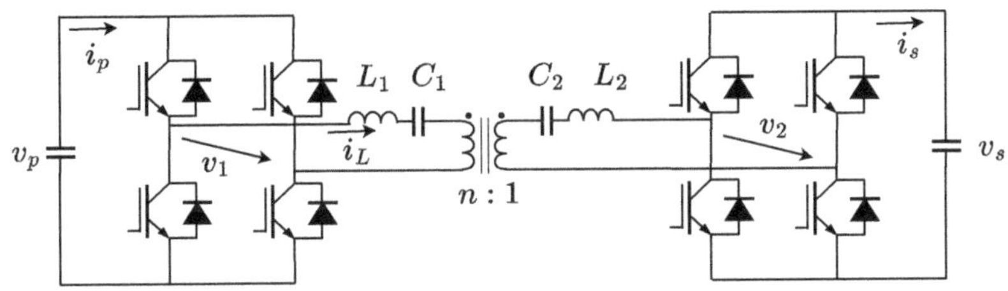

Figure 1. Topology of a resonant CLLC DAB converter.

In this figure, v_p, i_p, v_s and i_s are the input and output voltages and currents, respectively. L_1, C_1, L_2 and C_2 are the inductance and capacitance of the primary and secondary resonant tank, respectively. i_L is the primary current that circulates by inductance L_1. The voltages v_1 and v_2 are the alternating voltages of the primary H bridge and the secondary H bridge, respectively. Finally, n is the transform ratio between the primary and secondary.

The power converter can be modeled by a simple equivalent circuit, as depicted in Figure 2. For simplicity, no parasitic effects are considered (such as those that involve switches, inductance, or capacitance). Additionally, the magnetizing inductance is neglected.

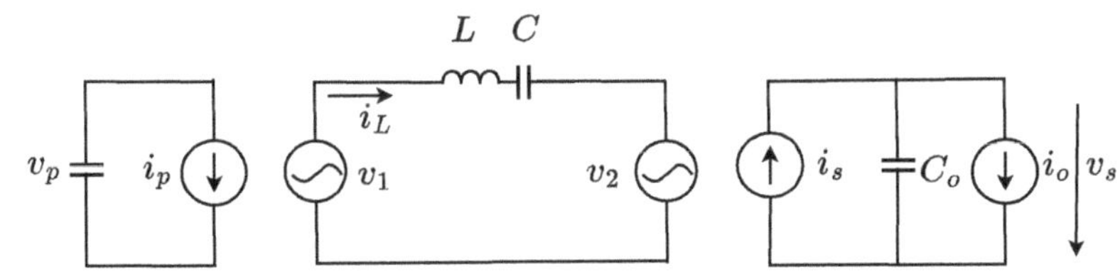

Figure 2. Simplified model of a resonant CLLC DAB converter with the input constant and output connected to resistance.

In this figure, L represents the equivalent inductance $L = L_1 + n^2 L_2$, and C is the equivalent capacitance $C = C_1 + n^2 C_2$.

Depending on the application, the input voltage v_p or the output voltage v_s can be continuously maintained by a power supply or can be connected to a current source. In this case, a current source connected to the output is considered and the input voltage is constantly maintained.

In terms of the Laplace transformation, this system can be considered as two subsystems, as shown in Figure 3: the resonant tank subsystem and the RC output filter subsystem. In this figure, the variables in capital letters V_1, V_2, I_L, I_s and I_o represent the phasor name of the same lowercase letter defined previously. s is the Laplace symbol.

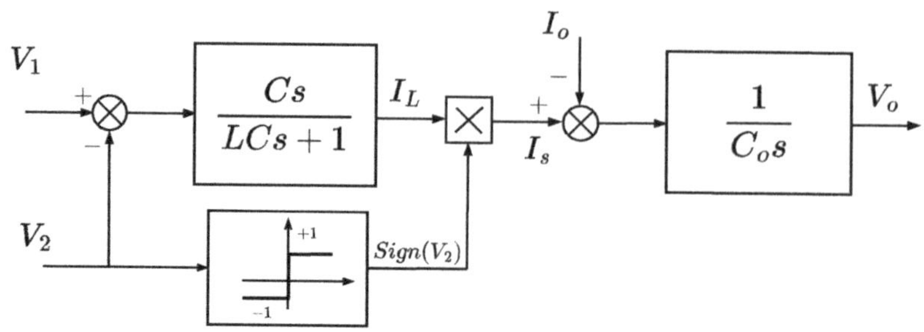

Figure 3. Simplified model based on transference function of a resonant CLLC DAB converter.

V_1 and V_2 are square waveforms shifted at ϕ angle. Using the first-harmonic approximation, the first harmonic of the difference between V_1 and V_2 can be obtained using the Fourier series. The first harmonic is obtained in Equation (1):

$$(V_1 - V_2)_1 = \frac{4\,\mathrm{V_s}\,n\,\sin(\phi)}{\pi} + \frac{4\,\cos(\phi)\,(\mathrm{V_p} - \mathrm{V_s}\,n)\,\mathrm{i}}{\pi}. \tag{1}$$

In this equation, V_p and V_s are the DC component of the input and output voltage, respectively. i is the imaginary unity complex number. The admittance is obtained as follows:

$$G = -\frac{C\,\omega\,\mathrm{i}}{C\,L\,\omega^2 - 1}. \tag{2}$$

The current I_L can be obtained by multiplying the admittance by the difference in voltages. The current I_L results are obtained as follows:

$$I_L = G(V_1 - V_2)_1 = \frac{4\,C\,\mathrm{V_s}\,n\,w\,\sin(\phi)}{\pi\,(C\,L\,w^2 - 1)} - \frac{4\,C\,w\,\cos(\phi)\,(\mathrm{V_p} - \mathrm{V_s}\,n)\,\mathrm{i}}{\pi\,(C\,L\,w^2 - 1)}. \tag{3}$$

Using this expression, the output current I_o can be obtained by projecting the current I_L into the output voltage per unit $V_2(pu)$. This $V_2(pu)$ voltage can be expressed by first harmonic approximation as follows:

$$V_2(pu) = \frac{2\,(\sin(\phi) - \cos(\phi)\,\mathrm{i})}{\pi}. \tag{4}$$

The output current I_s is a function of the phase angle ϕ and can be obtained, as explained above, using the scalar product of I_L by $V_2(pu)$. The result is as follows:

$$I_s = V_2(pu) \cdot I_L = \frac{8\,C\,\mathrm{V_p}\,\omega\,\sin(\phi)}{\pi^2\,(C\,L\,\omega^2 - 1)}. \tag{5}$$

This approximate model considers the dynamic decomposition of the fast dynamic and slow one. This can be exploited for the design of distributed cascade controllers so that the dynamics of the current controller can be much faster than that of the output voltage controller.

2.3. System Architecture and CLLC Resonant Converters

The proposed system focuses on the parallelization of CLLC resonant converters within hybrid AC–DC microgrids. Figure 4 shows a photo of the power converter as a modular system which has been used for our experiments. The parallel connection of these modules is represented in Figure 5. These converters are essential in solid-state transformers (SSTs), which enable bidirectional power transfer with high efficiency. A key advantage of CLLC converters is their ability to operate in high-frequency transformer configurations, which are typically used in modular power systems with power ratings of around 25 kW per module. To achieve higher power levels, multiple converters need to be operated in parallel at the output stage.

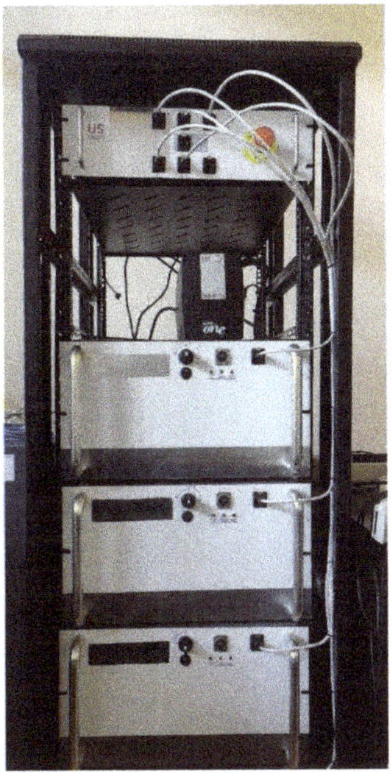

Figure 4. Parallel CLLC resonant converter system physical configuration.

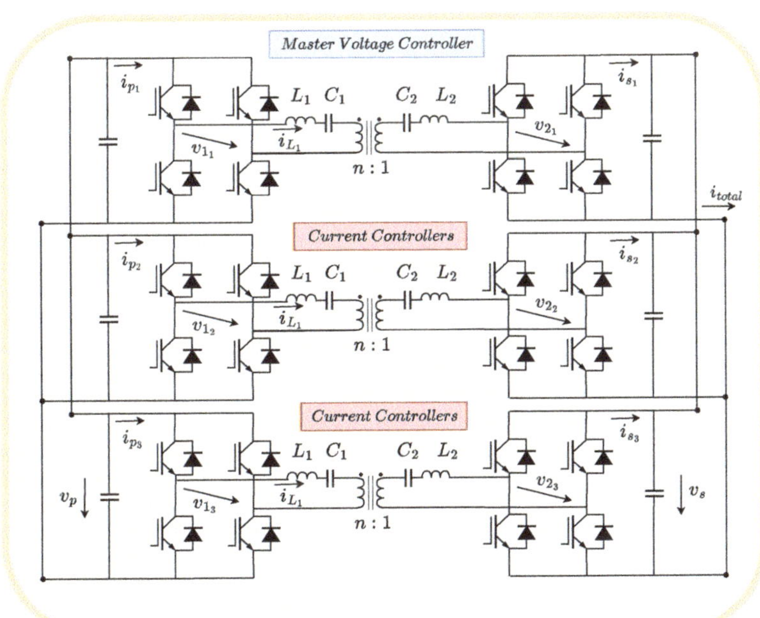

Figure 5. Structure of parallel resonant CLLC DAB converters.

2.4. Traditional Control Strategies

Traditional control strategies for parallel converters in SSTs typically rely on droop control, either based on voltage or current. In voltage droop control, the output voltage is adjusted based on load changes, while current droop control adjusts the current supplied by each converter depending on the system's load demand. However, these approaches face several challenges when applied to large-scale, modular systems. In parallel systems, droop control can lead to unequal current sharing between converters, resulting in circulating currents and efficiency losses. Systems controlled via global controllers (e.g., PLC-based systems) often exhibit slow response times due to communication delays, especially during fast load changes.

Hybrid control schemes that combine voltage and current droop strategies have been proposed to mitigate these issues but continue to suffer from transient instability and require complex coordination across multiple units.

2.5. Proposed Control Methodology

To overcome the limitations of traditional control methods, this paper introduces a distributed control strategy that eliminates the need for a global controller and instead leverages local control at the converter level. The overall structure of this approach is illustrated in Figure 6, showing the simplified model of the resonant CLLC DAB converter with constant input and a resistive load.

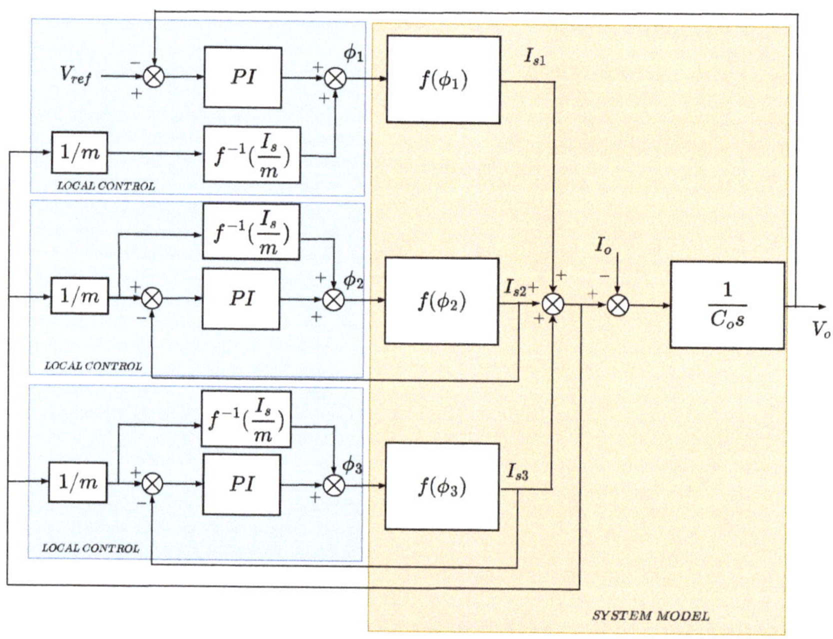

Figure 6. Distributed control system of DAB resonant converters.

2.5.1. Master Voltage Controller

As depicted in Figure 6, a master voltage PI controller is installed on one of the converters in the system. This controller is responsible for regulating the output voltage of the entire system, ensuring that the output voltage remains stable despite changes in load demand or fluctuations in input power. By localizing this control function to a single converter, the system avoids communication delays commonly associated with global controllers. This control strategy allows the system to emulate the behavior of an ideal transformer, where the voltage at the output can be dynamically adjusted to meet load requirements, regardless of the direction of power flow.

The master voltage controller operates by adjusting the output voltage reference for the system, allowing real-time regulation without relying on external communication links.

2.5.2. Distributed Current Controllers

Each remaining converter operates under a current PI control strategy (see Figure 6). The total load current is evenly distributed among the parallel converters, with each current controller regulating its assigned portion. This strategy ensures balanced current sharing and minimizes circulating currents, which are common in parallel converter systems.

The total load current is calculated by the system and divided equally between the active converters. Each converter is equipped with its own current PI controller, which adjusts its output current to match the reference value determined by the system. This distributed approach ensures that the converter is not overloaded and minimizes the risk of current imbalances.

As illustrated in Figure 6, the current controllers calculate their reference values by equally dividing the total i_{total} load current among the active converters m, allowing dynamic load sharing without the need for extensive communication between units.

2.5.3. Converter Characteristics

The CLLC resonant converter used in this study operates with the following nominal design parameters. Table 1 lists the nominal electrical specifications of the converter, and Table 2 provides the parameters of PI controllers.

Table 1. Nominal electrical specifications of the converter.

Parameter	Nominal Value
Input bus voltage (V_s)	550 V
Output bus voltage (V_p)	554 V
Input current (I_s)	45 A
Output current (I_p)	45 A
Output power (P_{out})	25 kW
Switching frequency (f_{sw})	80 kHz
Transformer turn ratio (n)	$n = \frac{N_{prim}}{N_{sec}} = 1.007$

Table 2. Nominal PI controller parameters.

Controller	K_p	K_i
Master Voltage Controller	1.2	5.7×10^{-6}
Distributed Current Controller	2.4	4×10^{-4}

These nominal parameters were carefully selected to ensure stable and efficient system operation under the experimental and simulation conditions described in this work.

The oscillation method, also known as the Ziegler–Nichols method, was used to adjust PI controllers for voltage and current regulation in CLLC converters through phase-shift control between primary and secondary bridges, with a fixed duty cycle of 50% on both sides. Resonant converters operate near their natural frequency, where the dynamics tend to be oscillatory, allowing for the direct application of the oscillation method to identify critical stability parameters. In this method, the proportional gain K_p is gradually increased until the system reaches the critical gain K_c and exhibits sustained oscillations with a well-defined oscillation period P_c. The proportional gain is set as $K_p = K_c$, and the integral constant K_i is calculated as

$$K_i = \frac{K_p}{T_i}, \quad \text{where} \quad T_i = \alpha P_c,$$

and α is a chosen fraction of P_c. The procedure is applied sequentially: first, the voltage PI controller is tuned to regulate the output voltage, followed by the current PI controller, ensuring that the conditions set in the voltage loop are maintained. This approach provides an intuitive, yet robust, tuning process, improving system stability and dynamic performance by enabling precise regulation of voltage and current. The method's effectiveness lies in its ability to optimize controller parameters without requiring complex mathematical models, enhancing the efficiency and control accuracy of the CLLC converter operating near its resonant frequency.

2.5.4. Feed-Forward Term in the Control Strategy

To further improve the transient response and enhance system stability, the proposed method incorporates feed-forward control into the current regulation scheme. Feed-forward control allows for faster responses to changes in load conditions by anticipating disturbances and adjusting the current output accordingly. This approach significantly reduces the reaction time of the system and prevents overshooting or instability during transient events.

By placing both voltage and current control functionalities locally within each converter, the system eliminates communication delays inherent in centralized control systems, making it more responsive to real-time changes in power demand and improving overall stability.

The feed-forward term in Equation (6) can be obtained by taking the inverse of Equation (5).

$$\phi_{\mathrm{ff}} = \arcsin\left(\frac{\pi^2\,(C\,L\,\omega^2 - 1)\,I_{total}}{8\,C\,V_{\mathrm{p}}\,\omega\,m}\right). \tag{6}$$

In Equation (6), m is the number of converters in parallel, and I_{total} is the total current that is measured by each converter.

2.6. *Algorithm for Voltage and Current Regulation*

The core of the proposed control strategy is a novel algorithm designed to manage both voltage regulation and current sharing. The algorithm is implemented across all converters, but its operation is divided between the master voltage controller and the individual current controllers.

2.6.1. Voltage Regulation Algorithm

The voltage regulation algorithm used by the master controller is based on continuous monitoring of input and output voltages. The difference between the output voltage and the input voltage, multiplied by the transformer ratio, is calculated in real-time. The controller then adjusts the output voltage to minimize this difference, ensuring that the system behaves as a transformer that can operate independently of the power flow direction.

2.6.2. Current Sharing Algorithm

The current sharing algorithm, implemented by the distributed current controllers, calculates the total load current and divides it evenly among the converters. The algorithm monitors the output current of each converter, comparing it with the reference current calculated from the total load. If discrepancies are detected, the current controller adjusts the output to ensure balanced sharing.

The inclusion of feedforward control in the current PI controller enhances this process by allowing the controllers to adjust the output pre-emptively when load changes are detected. This prevents sudden changes in the current distribution and ensures that the system remains stable under dynamic conditions.

3. Experimental Results

3.1. Experimental Setup

The experimental setup was designed to evaluate the performance of the proposed distributed control system for parallelized CLLC resonant converters within a hybrid AC–DC system. The test system included a combination of DC–DC converters, DC–AC converters, and a DC microgrid for load simulation and power distribution.

To ensure consistent operation and accurate evaluation, the parameters specified in Table 1 were applied during all experiments. These nominal values were carefully selected to match the requirements of the modular CLLC converters used in the tests.

The configuration is described as follows:

- A DC input bus was connected to a DC–AC converter. This converter allowed the DC bus to connect to the AC electrical grid, enabling bidirectional power flow and grid-tied operation.
- Multiple CLLC resonant DC–DC converters were connected in parallel. The converters shared the load current evenly and the proposed distributed control strategy was applied to regulate the sharing of voltage and current among these converters.
- The output of the DC–DC converters was connected to a DC microgrid, which represented the load of the system. This microgrid operated autonomously and was used to test the system's performance under varying load conditions.
- The DC microgrid was stabilized using a battery storage system. The battery served as a voltage regulator for the microgrid, ensuring stable operation and preventing large voltage fluctuations under dynamic load conditions. The battery controlled the microgrid voltage, allowing smooth power flow and load balance.

This configuration was used to evaluate the voltage regulation, current sharing, and transient response of the proposed control strategy, focusing on the following key performance metrics:

- System stability during normal and dynamic load conditions.
- Transient response time to sudden changes in load.
- Performance of current sharing between parallel DC–DC converters.

3.2. Test 1: Positive AC Power Flow

In the first test, we assign a series of positive AC power references to the DC-AC converter to simulate an increase in power demand. Several increments were applied, and the system response was recorded. The following results were obtained:

- Figure 7: Graph of the individual currents of each converter when the power reference is positive, showing the percentage of current distributed between the three converters. The graph confirms that the current sharing between the converters is well balanced at different power levels.
- Figure 8: Oscilloscope capture of the total current flowing through the DC output bus. This image highlights the stability of the total current during load increments.

The results indicate that under positive AC power flow, the system maintained balanced current sharing among the three converters, with minor deviations remaining within acceptable limits. For positive current flow, the system maintained deviations less than 3% at full load and low load. During transient events, the deviations were slightly larger but remained below 8.15%.

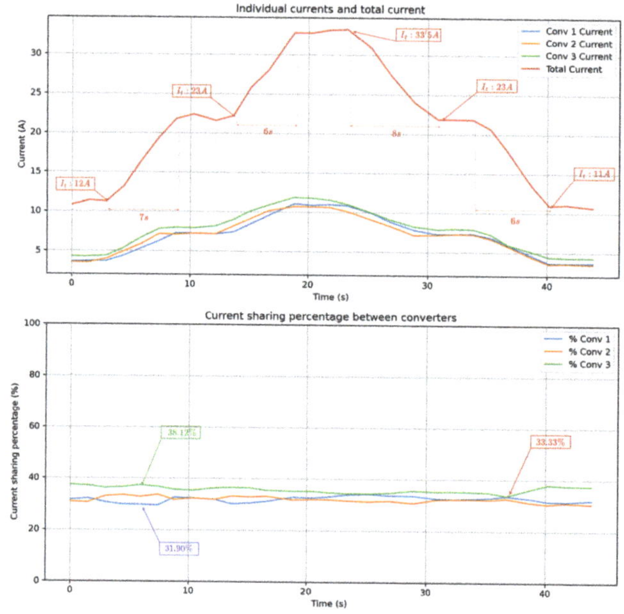

Figure 7. Individual currents of each converter and total current at positive power reference and percentage share of total current between the converters.

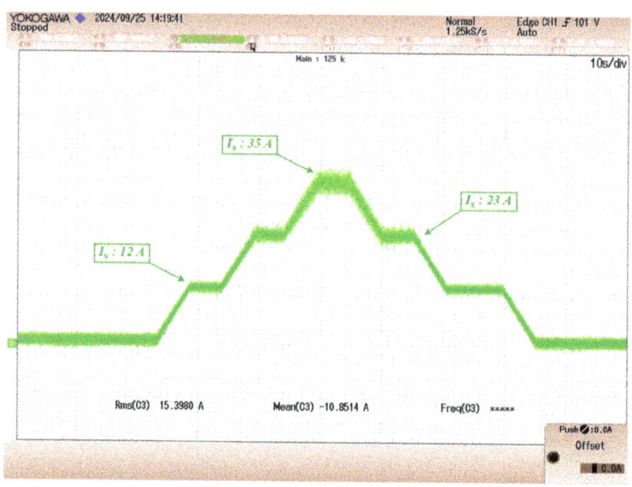

Figure 8. Positive variation of total current through the DC bus. Frequency is not detected in DC mode and its value appears as stars. Time scale: 10 A/div.

3.3. Test 2: Negative AC Power Flow

In the second test, the DC–AC converter was assigned negative AC power references, simulating sending power back to the grid. Again, multiple power increments were applied, and the system response was recorded. The following results were obtained:

- Figure 9: Plot of the individual currents of each converter when the power reference is negative, showing the percentage of current distributed among the three converters.

- Figure 10: Oscilloscope capture of the total current flowing through the DC output bus, highlighting the stability of the total current during power surges with negative power flow.

For negative AC power flow, the results showed a similar balanced current sharing between the converters, confirming the system's ability to effectively handle bidirectional power flow. For negative current flow, deviations were kept below 1% at full and low load and below 9.75% during transient events.

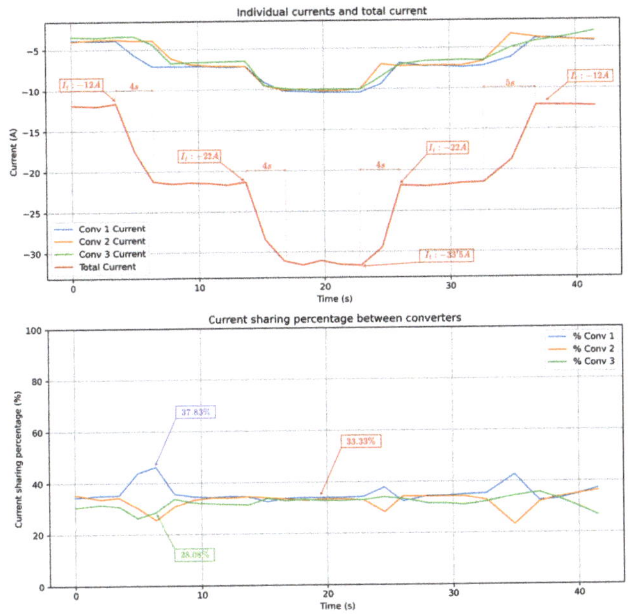

Figure 9. Individual currents of each converter and total current at negative power reference and percentage share of the total current between converters.

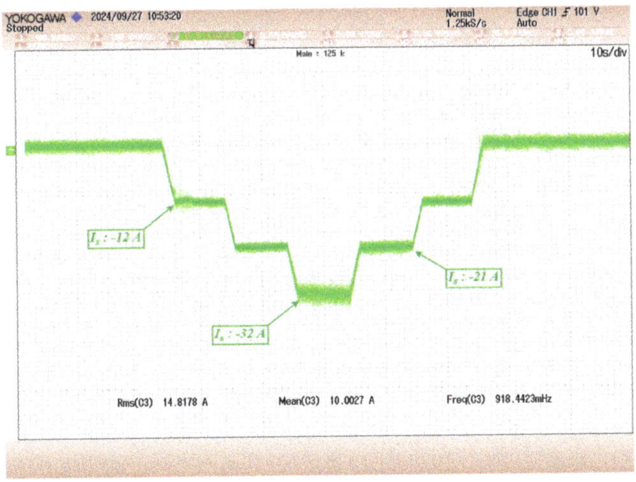

Figure 10. Negative variation of total current through the DC bus. Time scale: 10 A/div.

Summarizing the results, both tests performed with positive and negative power references confirmed the correct operation of the distributed control system for the parallelized CLLC converters. The key observations from the experiments are as follows:

- The current sharing between the converters remained within acceptable limits, with minimal deviations under all load conditions.
- The system maintained a stable power flow through the DC bus during both positive and negative power operations.
- The system responded smoothly to incremental changes in power demand, with stable voltage regulation and minimal fluctuations in current sharing.

The data from both tests are summarized in the consolidated Table 3, which shows the optimal and suboptimal results of each test in terms of the maximum difference between the percentages of sharing.

Table 3. Optimal and sub-optimal energy flow testing results.

Current (A) Conv 1	Current (A) Conv 2	Current (A) Conv 3	Total Current (A)	% Share Conv 1	% Share Conv 2	% Share Conv 3	Max Difference (%)
5.94	5.76	5.96	17.66	33.64	32.62	33.75	1.13
11.01	10.64	11.56	33.21	33.15	32.04	34.81	2.77
3.67	3.54	4.41	11.62	31.58	30.46	37.95	7.49
3.64	3.42	4.35	11.41	31.90	29.97	38.12	8.15
−3.86	−3.95	−3.90	−11.71	32.96	33.73	33.30	0.77
−10.23	−10.15	−9.93	−30.31	33.75	33.49	32.76	0.99
−3.90	−4.02	−3.45	−11.37	34.30	35.36	30.34	5.01
−9.39	−6.97	−8.46	−24.82	37.83	28.08	34.09	9.75

3.4. Test 3: System Response to Bidirectional Nominal Power Steps

This test scenario describes a step response test to evaluate the performance of the system's power regulation. Applying a power step from −24 kW to +24 kW (a jump of 48 kW) tests the system's ability to manage a rapid change in power demand. This shift corresponds to a current change of 87 A, challenging both the current control loop and the voltage stability mechanisms across converters and buses.

The test setup allows us to observe the following key response characteristics.

Current Response on the Internal Bus: Monitoring the total current through the internal bus provides insight into the dynamic response of the system. Specifically, observing the current rise, peak, and any oscillations or settling times helps assess whether the current control loop is appropriately tuned.

Voltage Stability Across Converter Buses: Monitoring the voltage across buses on either side of the converters reveals how effectively the converters manage energy during power transients. Stability in controlled voltage (Vs) is observed, with maximum deviations of 3 V, indicating an effective response of the system. Voltage spikes, drops, or prolonged settling times would signal the need to adjust converter control parameters or enhance energy storage/buffer capacity.

In this context, Figure 11 typically illustrates these dynamics, showing the following:

Rapid response of the total current in the internal bus, highlighting the overshoot and settling time following the power step. Voltage trends on both sides of the converters to verify that they remain within safe operating limits and promptly return to steady-state values after the step change, with controlled voltage (Vs) deviations remaining within 3 V. These data support diagnostics on the system's robustness in handling significant power variations, ensuring stability and reliability under diverse operating conditions.

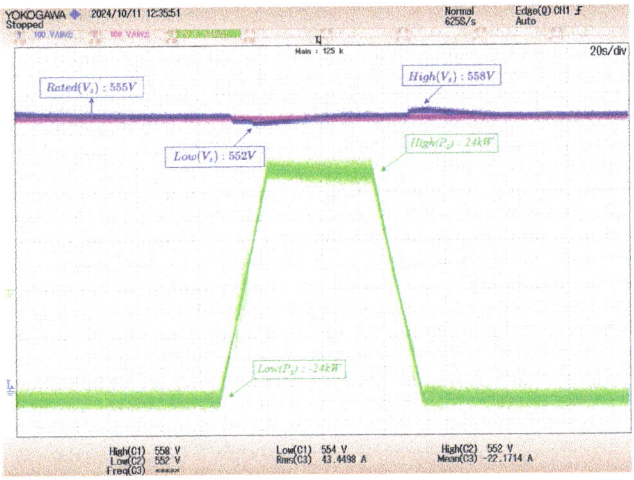

Figure 11. Graph showing the response to changes in reference power. The pink trace represents the external bus voltage, the blue trace represents the internal bus voltage, and the green trace represents the total current on the internal DC bus. Frequency is not detected in DC mode and its value appears as stars. Voltage scale: 100 V/div. Current scale: 20 A/div.

3.5. *Test 4: System Response to Voltage Variations with and Without Power Reference*

This test is designed to evaluate how the system responds to external bus voltage variations, both under load and nonload conditions. Four scenarios have been simulated, with changes in the external bus voltage of 20 V, both increases and decreases, and the behavior of the current control has been evaluated in two situations: with a power reference of 10 kW and without a power reference (0 W). The main objective has been to analyze the speed and stability of the system response under different conditions.

- Figure 12: Voltage rise from 543 V to 563 V without power reference (0 W). In this scenario, the system was subjected to an increase in the voltage on the external bus, from 543 V to 563 V, without applying any power reference. This first scenario evaluated how the current control performs when the system operates with no load and only focuses on stabilizing the internal voltage. Here, the speed with which the system balances the voltages, as well as the stability during the transition, were the main points to be analyzed.
- Figure 13: The voltage rises from 543 to 563 V with a power reference (10 kW). In this second scenario, the same 20 V voltage rise was repeated on the external bus, but this time, a power reference of 10 kW was maintained. The test allowed us to observe how the current control adjusts power sharing and current management in parallel connected converters under active load conditions. The interaction between voltage stabilization and power maintenance was key in this analysis.
- Figure 14: Voltage drop from 563 V to 543 V without power reference (0 W). In the third scenario, a 20 V voltage drop from 563 V to 543 V was simulated without a power reference (0 W). This scenario was designed to test how the system restores the voltage balance when the grid experiences a drop without the system operating under load. The speed of the reaction and the absence of oscillations in the internal bus are indicators of the efficiency of the control.
- Figure 15: Voltage drop from 563 V to 543 V with power reference (10 kW). Finally, in the fourth scenario, the 20 V voltage drop was repeated while keeping the power reference at 10 kW. This scenario tested the ability of the current control to simul-

taneously manage voltage stabilization and power sharing between the converters, keeping the load active without interruptions or significant fluctuations.

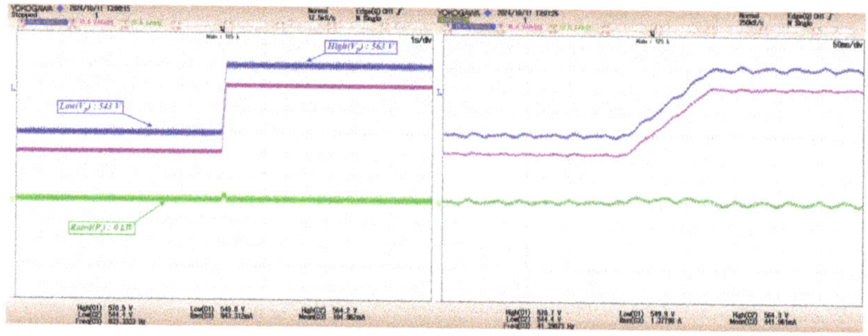

Figure 12. Evolution of the external bus voltage during the rise from 543 V to 563 V without power reference. The pink trace represents the external bus voltage, the blue trace represents the internal bus voltage, and the green trace represents the total current on the internal DC bus. Voltage scale: 10 V/div. Current scale: 20 A/div.

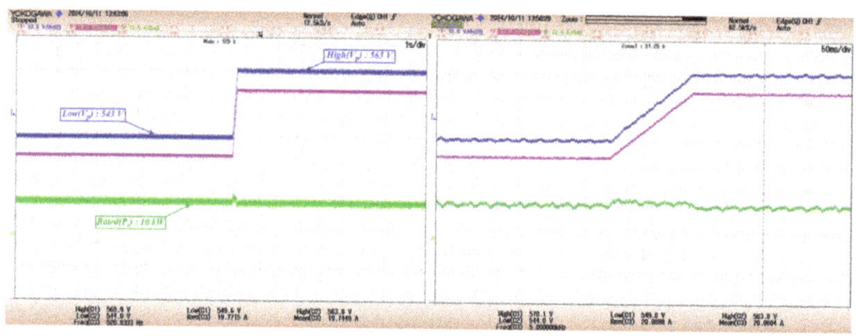

Figure 13. Evolution of the external bus voltage during the increase from 543 V to 563 V with power reference at 10 kW. The pink trace represents the external bus voltage, the blue trace represents the internal bus voltage, and the green trace represents the total current on the internal DC bus. Voltage scale: 10 V/div. Current scale: 20 A/div.

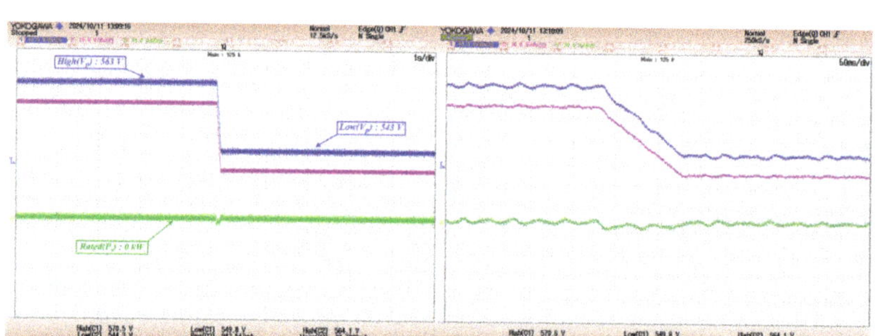

Figure 14. Evolution of the external bus voltage during the drop from 563 V to 543 V without power reference. The pink trace represents the external bus voltage, the blue trace represents the internal bus voltage, and the green trace represents the total current on the internal DC bus. Voltage scale: 10 V/div. Current scale: 20 A/div.

In the four scenarios evaluated, the system showed robust and stable behavior, responding quickly to external bus voltage changes, both during 20 V rises and falls. In all cases, the current control acted efficiently to stabilize the internal bus and ensure a smooth transition without significant fluctuations. The current distribution between the converters was balanced, which allowed system stability to be maintained throughout the voltage variations, ensuring continuous and reliable operation.

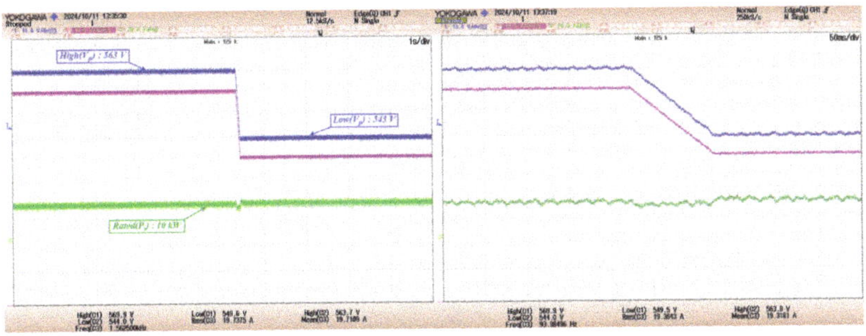

Figure 15. Evolution of the external bus voltage during the step-down from 563 V to 543 V with power reference at 10 kW. The pink trace represents the external bus voltage, the blue trace represents the internal bus voltage, and the green trace represents the total current on the internal DC bus. Voltage scale: 10 V/div. Current scale: 20 A/div.

3.6. Test 5: Evaluating the Effect of Feed-Forward on System Control

In this test, the aim is to evaluate the influence of the feed-forward term in the control system in order to determine whether its incorporation improves its performance. To this end, a comparison will be made between the results obtained when carrying out the test with and without the feed-forward term, which will allow the reaction times of the control to be analyzed in each case, as shown in Figures 16 and 17.

A current step has been applied, increasing from 4 A to 14 A, to observe the effect of this term on the dynamics of the system. This configuration will make it possible to evaluate the effectiveness of the feed-forward and its capacity to improve the response of the control to changes in the current.

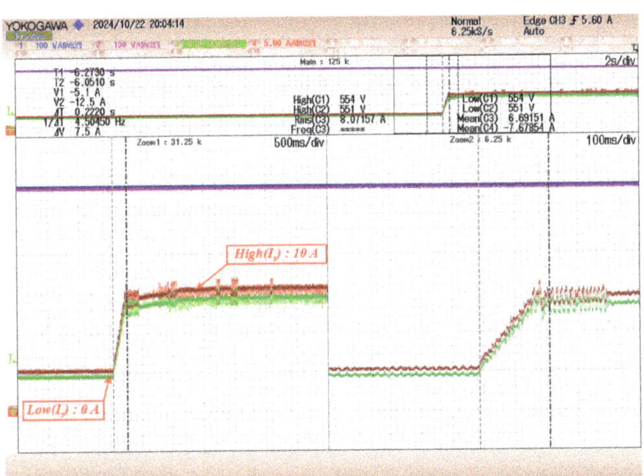

Figure 16. Current behavior without the feed-forward term at 10 A step. The pink trace represents the external bus voltage, the blue trace represents the internal bus voltage, the green trace represents the total current on the internal DC bus, and the red trace represents the individual voltage for a single unit. Frequency is not detected in DC mode and its value appears as stars. Voltage scale: 100 V/div. Current scale: 5 A/div.

The results show that the inclusion of the feed-forward term allows a significantly faster system response. Without feed-forward, the time to reach the reference was 0.222 s, while with feed-forward, the reaction time was reduced to 0.046 s. This difference highlights the effectiveness of feed-forward in improving control behavior to changes in current.

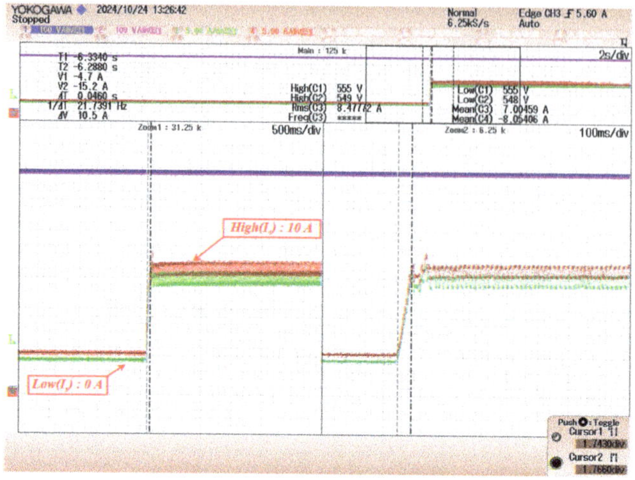

Figure 17. Current behavior with feed-forward term at 10 A step. The pink trace represents the external bus voltage, the blue trace represents the internal bus voltage, the green trace represents the total current on the internal DC bus, and the red trace represents the individual voltage for a single unit. Frequency is not detected in DC mode and its value appears as stars. Voltage scale: 100 V/div. Current scale: 5 A/div.

4. Discussion

The experimental results presented in the previous section validate the effectiveness of the proposed distributed control strategy for parallelized CLLC resonant converters. In this section, we discuss the significance of these results, the advantages of the proposed system over traditional control systems, and the potential impact of implementing external PLC-based control. Key performance metrics, such as response time, stability, and current sharing, are also addressed.

4.1. Improved Response Time and Stability

One of the most significant advantages of the proposed distributed control system is the fast response time and improved stability observed during both positive and negative power flow tests. The system was able to respond to dynamic load changes within 20 ms, significantly reducing transient time compared to traditional centralized control approaches, where communication delays often result in response times exceeding 150 ms.

The fast response time is largely due to the integration of local controllers within each converter, allowing real-time adjustments to current and voltage without relying on a central control unit. In contrast, a PLC-based system, where control decisions are made externally, would introduce communication delays that slow the system's reaction to load changes. This delay can lead to instability during transient events, particularly when handling sudden changes in the direction of the load or power flow.

In the proposed system, the inclusion of feed-forward control further enhances the dynamic response by allowing the current controllers to anticipate disturbances and adjust their outputs pre-emptively. This proactive adjustment improves system stability during transitions, ensuring smooth operation even under challenging conditions, such as rapid load fluctuations.

4.2. Current Sharing and Load Balancing

Experimental results also confirmed that the system maintained a balanced current sharing between all converters in parallel, with deviations of less than 0.8% during negative current flow at full load and low load conditions. In transient events, deviations remained below 9.75% for all converters. For positive current flow, the system showed slightly larger deviations, with a deviation of 3% at both full and low load and deviations of up to 8.15% during transient conditions. These deviations, although somewhat larger than those of the negative current flow, remained within acceptable limits for safe and efficient operation.

The precise current-sharing performance of the system is a result of the distributed current control implemented within each converter. This distributed control allowed the system to divide the total load current equally among the converters without requiring continuous communication between them. In contrast, a centralized PLC-based control system would likely struggle to maintain such a high level of current sharing accuracy due to communication delays and the inherent limitations of a centralized architecture.

In a PLC-based centralized control system, current imbalances would be more likely, as the central controller would need to gather data from each converter, process them, and send appropriate control signals. These delays can cause circulating currents and uneven load distribution, particularly during rapid load changes, leading to reduced system efficiency and higher operational risks.

The distributed control system, by placing control responsibility at the converter level, effectively mitigates these issues. It ensures balanced current sharing even under dynamic conditions, without the need for extensive communication between converters, enhancing both efficiency and reliability.

4.3. Impact of Control Strategy on Power Flow

The system's ability to handle bidirectional power flow effectively was demonstrated in both positive and negative AC power flow tests. This capability is essential for applications in hybrid AC–DC microgrids, where the direction of the power flow can change depending on the needs of the grid, for example, when feeding energy back into the grid or storing excess energy in a battery system.

The proposed voltage regulation algorithm, which continuously monitors and adjusts the output voltage, allowed the system to behave like an ideal transformer. This ensured that the system could regulate both the input and output voltage regardless of the direction of the power flow, maintaining stability and minimizing voltage fluctuations.

In contrast, a PLC-based control system would likely face challenges in maintaining stable voltage regulation during power flow reversal. The delays introduced by communication between the converters and the central controller could result in voltage oscillations and instability, particularly during transitions between power flow directions. These challenges are exacerbated in high-power systems, where even small voltage deviations can lead to significant inefficiencies and operational risks.

The results of the experiments confirm that the distributed control system offers superior performance in handling bidirectional power flow, making it a robust solution for applications in microgrids, renewable energy systems, and energy storage solutions.

4.4. Benefits of Local Control vs. External PLC-Based Control

The experimental results clearly show that the use of local control provides substantial benefits over traditional PLC-based control systems, particularly in terms of the following:

- Response time: Local control allows instantaneous adjustments to voltage and current without relying on external communication, which is crucial for maintaining system stability during transient events.

- Scalability: The plug-and-play nature of the distributed control system enables easy integration of new converters, while centralized control systems require significant reconfiguration and recalibration when expanding the system. With the proposed configuration, the central power plant controller is not affected by the number of units that are connected in parallel, and so this solution is more flexible and scalable if compared with centralized controller. Therefore, the proposed configuration is suitable for applications such as hybrid networks and microgrid applications, where any upgrade can be easily made, and a higher number of units in parallel is required.
- Robustness: By distributing control responsibility among individual converters, the system becomes more robust against failures. If one controller fails, the rest of the system can continue to operate without significant disruption. In a PLC-based system, the failure of the central controller could lead to a complete system shutdown.

In a scenario where external PLC control is used, the system would be limited by the communication latency between the converters and the PLC. These delays, although acceptable for some low-frequency applications, become problematic in high-frequency, high-power systems, such as those required for solid-state transformers (SSTs). In such systems, even small delays can lead to voltage instabilities and inefficient current sharing, increasing the risk of circulating currents and reducing overall system efficiency.

4.5. Comparison of Distributed and Centralized Control Methods in DC Microgrids

To evaluate the advantages of distributed control over traditional centralized methods, a comparison is made in four key aspects: system stabilization, current balancing, dynamic response to load and power variations, and computational complexity. The experimental results from [17] serve as a reference for this analysis.

It is important to highlight that traditional droop control presents significant limitations in DC microgrids. This method can lead to voltage deviations under heavy load conditions and inefficient power sharing during light load situations. Furthermore, the lack of proper regulation can result in system instabilities, especially in the presence of non-linear loads. These shortcomings make traditional droop control less effective in applications where high precision and stability in power supply are required [18].

In terms of stabilization, centralized control systems, such as those based on PLCs analyzed in [17], exhibit significant limitations due to their low sampling frequency, which is limited to 250 Hz. This frequency restricts how quickly the controller can adjust system variables, leading to delays in both current and voltage stabilization. Additionally, communication delays between the PLC and the local converters further exacerbate these limitations, as control commands and measurements must traverse the network, introducing latency [19]. In contrast, the distributed control proposed in this work leverages local PI controllers implemented in high-speed DSPs with a sampling frequency of 10 kHz. The key advantage of the decentralized system is that both the control commands and the measurements originate locally within each converter, eliminating communication delays and enabling a significantly faster response. The use of these DSPs, combined with data acquisition through high-frequency ADCs, ensures rapid adjustments and enhances overall system performance.

Current balancing is another critical aspect of parallel converters. In centralized systems utilizing droop control, as analyzed in [17], current imbalances are observed, especially under dynamic conditions. Experimental results show current deviations reaching up to 5–8%, particularly during load changes. Moreover, in the proposed centralized control, a low-pass filter (LPF) is included, which destabilizes the system below its cutoff frequency, and with low time constants, oscillations in the signal are produced, further aggravating stability issues.

In contrast, the proposed distributed control, with local PI controllers and a feed-forward term, anticipates load changes and precisely adjusts the converter output. This predictive capability ensures a robust system response, even during sudden power variations of −24 kW to +24 kW, maintaining voltage stability with deviations below 3 V. Additionally, the distributed system guarantees equitable current sharing between converters, keeping deviations below 3% in both steady-state and dynamic conditions. Current distribution eliminates unwanted oscillations and minimizes circulating currents.

The proposed distributed control method also reduces computational complexity compared to centralized approaches. In centralized systems, the computational load increases linearly with the number of converters, as a central controller must aggregate real-time data, solve global control equations, and dispatch control actions. This process becomes increasingly expensive and introduces delays as the system scales.

In contrast, the decentralized approach distributes control tasks among local PI controllers, which independently perform lightweight real-time computations for current regulation on digital signal processors (DSPs). The master voltage controller, responsible for the regulation of the output voltage, requires minimal computational resources due to the simplicity of PI operation. By eliminating the need for a central controller to process global system states, the proposed method ensures that computational resources scale efficiently with the number of converters, enhancing both system scalability and real-time performance.

5. Conclusions

This study presents a novel distributed control strategy for parallelizing CLLC resonant converters in hybrid AC-DC microgrids, addressing challenges like current sharing, voltage regulation, and system stability. The proposed approach eliminates the reliance on centralized control systems, replacing them with local controllers that allow faster and more accurate responses to changes in load conditions.

The experimental results demonstrated the following key findings:

- Improved Response Time: The distributed control system achieved significantly faster transient response times (15–20 ms) compared to the expected delays in PLC-based systems. The inclusion of feed-forward control further enhanced the system's ability to anticipate and react to disturbances, maintaining system stability under both positive and negative power flow conditions.
- Balanced Current Sharing: The system maintained a balanced current distribution among the parallel converters, with deviations kept below 1% for negative current flow at full load and low load. For positive current flow, deviations were 3% at both full and low load. During transient events, the deviations remained below 8% for negative currents and below 8.15% for positive currents. This accurate current-sharing was achieved without the need for complex communication between converters, relying on local current controllers to manage load sharing autonomously.
- Scalability and Flexibility: The proposed distributed control strategy inherently provides scalability due to its modular architecture. Each converter operates independently with local controllers, enabling the seamless integration of additional converters to meet higher power demands without requiring a major reconfiguration of the system. This modularity ensures that the system can expand easily while maintaining stable and balanced operation. Furthermore, the system demonstrates flexibility in handling varying load conditions and bidirectional power flow. The experimental results confirm that the system maintains dynamic stability under sudden power reference changes and external voltage variations, highlighting its adaptability to dynamic operating scenarios.

- Voltage Stability: The system's voltage regulation algorithm, which adjusts the output voltage in real-time based on the difference between the input and output voltages (scaled by the transformer ratio), ensured stable operation in all test scenarios. The system behaved like an ideal transformer, with minimal voltage fluctuations during both positive and negative power flows.

By distributing control responsibilities among individual converters, the system became more robust and adaptable, allowing it to continue operating even if one controller failed. This plug-and-play capability further enhances the system's practical applicability, particularly in scenarios where fast response and high reliability are essential.

Future Work: Looking ahead, further research could focus on incorporating artificial intelligence (AI) or machine learning to enhance the control system's ability to adapt to more complex conditions and predict potential problems, such as load variation and deviations of the parameters. Furthermore, the use of advanced communication protocols could improve coordination between distributed controllers, enabling even more precise current sharing and voltage regulation in larger-scale systems. Although the proposed distributed control strategy demonstrates significant improvements in current sharing, voltage regulation, and transient response, stability remains a critical factor in systems with parallel converters. Studies such as [20] have shown that interactions between impedance and control can lead to low-frequency oscillations (LFO) if not addressed properly.

In our work, the careful tuning of PI controller parameters and the inclusion of feedforward control help mitigate these potential instabilities. The experimental results confirm stable operation under load variations and bidirectional power flow conditions. Nevertheless, for future studies, we aim to incorporate a more detailed stability analysis based on phase margin criteria and impedance-based methods to further enhance the system's robustness.

Author Contributions: Conceptualization, D.C.V.; methodology, D.C.V.; software, D.C.V. and A.M.C.; validation, D.C.V. and A.M.C.; formal analysis, D.C.V., A.M.C. and E.G.D.; investigation, D.C.V.; resources, D.C.V.; writing—original draft preparation, D.C.V.; writing—review and editing, visualization, supervision, project administration, funding acquisition, E.G.D., J.M.C.S. and F.R.R. All authors have read and agreed to the published version of the manuscript.

Funding: This work was supported by the Spanish State Research Agency (AEI) through the Project TED2021-131604B-I00 funded by MCIN/AEI/10.13039/501100011033 and by the European Union Next GenerationEU/ PRTR.

Data Availability Statement: The original contributions presented in the study are included in the article. Further inquiries can be directed to the corresponding author.

Acknowledgments: We extend our heartfelt gratitude to ENGREEN—Laboratory of Engineering for Energy and Environmental Sustainability at the University of Seville for their essential support in the completion of this research paper. Special thanks are also given to the members of the PowerUS research group for their invaluable contributions and collaboration at each stage of this work. During the preparation of this manuscript, the authors used CHATGPT V4 for the purposes of preparing the reference list during the state of the art compilation and to improve English writing. The authors have reviewed and edited the output and take full responsibility for the content of this publication.

Conflicts of Interest: The authors declare no conflicts of interest.

Abbreviations

The following abbreviations are used in this manuscript:

AC	Alternating Current
DC	Direct Current
DAB	Dual Active Bridge

DERs	Distributed Energy Resources
MGs	Microgrids
PLCs	Programmable Logic Controllers

References

1. Lasseter, R.H. Microgrids. In Proceedings of the Power Engineering Society Winter Meeting, New York, NY, USA, 27–31 January 2002; Volume 1, pp. 305–308.
2. Guerrero, J.M.; Vasquez, J.C.; Matas, J.; De Vicuña, L.G.; Castilla, M. Hierarchical control of droop-controlled AC and DC microgrids—A general approach toward standardization. *IEEE Trans. Ind. Electron.* **2011**, *58*, 158–172. [CrossRef]
3. Olivares, D.E.; Mehrizi-Sani, A.; Etemadi, A.H.; Canizares, C.A.; Iravani, R.; Kazerani, M.; Hajimiragha, A.; Gomis-Bellmunt, O.; Saeedifard, M.; Palma-Behnke, R.; et al. Trends in microgrid control. *IEEE Trans. Smart Grid.* **2014**, *5*, 1905–1919. [CrossRef]
4. Olivares, D.E.; Canizares, C.A.; Kazerani, M. A centralized energy management system for isolated microgrids. *IEEE Trans. Smart Grid.* **2015**, *5*, 1864–1875. [CrossRef]
5. Farhangi, H. The path of the smart grid. *IEEE Power Energy Mag.* **2010**, *8*, 18–28. [CrossRef]
6. Bidram, A.; Davoudi, A. Hierarchical structure of microgrids control system. *IEEE Trans. Smart Grid.* **2012**, *3*, 1963–1976. [CrossRef]
7. Yazdanian, M.; Mehrizi-Sani, A. Distributed Control Techniques in Microgrids *IEEE Trans. Smart Grid.* **2014**, *5*, 2901–2909. [CrossRef]
8. Dragicevic, T.; Sucic, S.; Vasquez, J.C.; Guerrero, J.M. DC microgrids—Part I: A review of control strategies and stabilization techniques. *IEEE Trans. Power Electron.* **2015**, *31*, 4876–4891. [CrossRef]
9. Li, L.; Xu, G.; Sha, D.; Liu, Y.; Sun, Y.; Su, M. Review of Dual-Active-Bridge Converters With Topological Modifications. *IEEE Trans. Power Electron.* **2023**, *38*, 9046–9076. [CrossRef]
10. Arshadi, S.A.; Ordonez, M.; Eberle, W. Current-Sharing Worst-Case Analysis of Three-Phase CLLC Resonant Converters. *IEEE Trans. Power Electron.* **2022**, *37*, 3099–3110. [CrossRef]
11. Gao, Y.; Wang, S.; Dragicevic, T.; Wheeler, P.; Zanchetta, P. Artificial Intelligence Techniques for Enhancing the Performance of Controllers in Power Converter-Based Systems—An Overview. *IEEE Open J. Ind. App.* **2023**, *4*, 366–375. [CrossRef]
12. Kim, J.-H.; Kim, C.-E.; Kim, J.-K.; Lee, J.-B.; Moon, G.-W. Analysis on Load-Adaptive Phase-Shift Control for High Efficiency Full-Bridge LLC Resonant Converter Under Light-Load Conditions. *IEEE Trans. Power Electron.* **2016**, *31*, 4942–4955. [CrossRef]
13. Zhou, K.; Sun, Y. Research on Bidirectional Isolated Charging System Based on Resonant Converter. *Electronics* **2022**, *11*, 3625. [CrossRef]
14. Zhu, T.; Zhuo, F.; Zhao, F.; Wang, F.; Yi, H.; Zhao, T. Optimization of Extended Phase-Shift Control for Full-Bridge CLLC Resonant Converter With Improved Light-Load Efficiency. *IEEE Trans. Power Electron.* **2020**, *35*, 11129–11142. [CrossRef]
15. Chen, H.; Sun, K.; Shi, H.; Wang, L.; Zhang, K. An Inner Phase Shift Control Scheme for the CLLC Converter. In Proceedings of the 2022 International Power Electronics Conference (IPEC-Himeji 2022—ECCE Asia), Himeji, Japan, 15–19 May 2022; pp. 1059–1064. [CrossRef]
16. Shakib, S.M.S.I.; Mekhilef, S.; Nakaoka, M. Dual Bridge LLC Resonant Converter with Frequency Adaptive Phase-Shift Modulation Control for Wide Voltage Gain Range. In Proceedings of the 2017 IEEE Energy Conversion Congress and Exposition (ECCE), Cincinnati, OH, USA, 1–5 October 2017; pp. 2265–2271. [CrossRef]
17. Silva, R.A.G.; Vilela, J.A., Jr. Experimental Investigation of Droop Control for Power Sharing of Parallel DC–DC Converters in Voltage and Current Mode Control. *J. Control Autom. Electr. Syst.* **2024**, *35*, 1008–1018. [CrossRef]
18. Srivastava, S.; Roy, L.H.; Ravi, G.; Sahay, K.; Tripathi, P.; Dongre, S. Droop Control based Control technique and Advancements for Microgrid Stability—A Review. In Proceedings of the 2023 IEEE Renewable Energy and Sustainable E-Mobility Conference (RESEM), Bhopal, India, 17–18 May 2023; pp. 1–6. [CrossRef]
19. Baros, D.; Rigogiannis, N.; Papanikolaou, N.; Loupis, M. Investigation of communication delay impact on DC microgrids with adaptive droop control. In Proceedings of the 2020 22nd European Conference on Power Electronics and Applications (EPE'20 ECCE Europe), Banja Luka, Bosnia and Herzegovina, 4–6 November 2020; pp. 1–10.
20. Lin, C.; Ge, X.; Wang, H.; Meng, X.; Mu, J.; Liu, Z. Low-Frequency Stability Analysis of Train-Grid System: A Perspective From DAB Converter. *IEEE Trans. Power Electron.* **2024**, *39*, 9066–9071. [CrossRef]

electronics

Article

FPGA Implementation of Nonlinear Model Predictive Control for a Boost Converter with a Partially Saturating Inductor

Alessandro Ravera, Alberto Oliveri *, Matteo Lodi and Marco Storace

Department of Electrical, Electronic, Telecommunications Engineering and Naval Architecture, University of Genoa, Via Opera Pia 11a, 16145 Genova, Italy; alessandro.ravera@edu.unige.it (A.R.); matteo.lodi@unige.it (M.L.); marco.storace@unige.it (M.S.)

* Correspondence: alberto.oliveri@unige.it

Abstract: Enhancing power density is a primary objective in electronic power converters. This can be accomplished by employing smaller inductors operating in partial magnetic saturation. In this study, an embedded digital controller is proposed, based on nonlinear model predictive control (NMPC), for the regulation of a DC–DC boost converter, exploiting a partially saturating inductor. The NMPC prediction model exploits a behavioral inductor model that accounts for magnetic saturation and losses and allows the converter regulation while enforcing constraints. The NMPC controller is implemented on a field programmable gate array (FPGA), demonstrating its real-time feasibility while successfully controlling a boost converter operating at switching frequencies up to 80 kHz. Hardware–software co-simulation results show accurate voltage regulation and constraint satisfaction, even under partial magnetic saturation.

Keywords: magnetic saturation; switched mode power supply; nonlinear model predictive control

1. Introduction

The magnetic saturation of inductor cores leads to a reduction in the material's permeability as the magnetic field within the core increases. From the perspective of the inductor, this results in a decrease in inductance as current increases—either gradually (as in iron powder core inductors) or abruptly (as in ferrite core inductors). Research has demonstrated that utilizing partially saturated inductors can enhance the power density of power converters, albeit with a slight increase in power loss. The groundbreaking work in [1] illustrates how inductor saturation can be managed in power converters with minimal impact on power consumption. Further studies, including [2,3], provide a more detailed analysis, also considering temperature effects [4].

Designing, simulating, and controlling power converters with partially saturating inductors requires accurate models to predict their operational behavior. Several nonlinear behavioral inductor models are reviewed in [5–7]. The range of validity of these models has been extended in [8–10] where inductance and power losses are reproduced for different operating frequencies, applied waveforms, air-gap lengths, and core materials.

Model predictive control (MPC), which can be easily implemented on digital circuits [11], is frequently applied to power converters [12,13], offering superior performance compared to traditional model-free proportional–integral (PI) regulators. MPC allows enforcing input and state constraints, particularly important for current limitations

https://doi.org/10.3390/electronics14050941

(for safety reasons) in power converters. Applications include four-switch, three-phase rectifiers in balanced grids [14] and inverters in unbalanced grids [15,16]. For boost converters, linear MPC is applied in [17], and nonlinear MPC (NMPC) in [18].

Despite these advancements, few implementations consider nonlinear inductance. Various nonlinearities, including magnetic saturation, are addressed in [19], where the inductance of a powder iron core inductor is modeled using an exponential function, and the MPC problem is solved with a fast gradient algorithm, which does not allow imposing current constraints. For ferrite core inductors, ref. [20] introduces an explicit linear MPC controller for a buck converter, relying on a simplified inductor model with step-like inductance and no losses.

In a previous study [21], we exploited nonlinear MPC for the voltage regulation of a DC–DC boost converter by imposing current constraints and using a ferrite core inductor model [9], which represents inductance as an arctangent function of current while accounting for instantaneous losses. This model provides significantly greater accuracy than the simplified approach in [20]. The main limitation of [21] is that only simulation results are proposed, without checking if the technique can be applied to a power converter in real time.

The main novelty of this paper is the FPGA implementation of the technique proposed in [21], which relies on a nonlinear inductor model, for obtaining an embedded real-time controller. This requires the use of limited hardware resources with tight constraints on the circuit latency. Proper algorithms should be therefore adopted for both the nonlinear programming and the numerical integration of the system for prediction. Moreover, fixed-point data representation is mandatory to fulfill real-time constraints.

Several FPGA-based linear MPC implementations are surveyed in [22]. Concerning NMPC on FPGA, different algorithms have been applied, including particle swarm optimization [23,24], mesh adaptive direct search (MADS) [25], and gradient-based techniques [26–28]. To perform the optimizations required by the NMPC approach, we chose the MADS algorithm, a zero-order method that does not require the evaluation of derivatives of the cost function. As shown in [25,29,30], it is particularly suitable for micro-controller and FPGA implementations, especially for small-size problems, as considered in this work. Other optimization algorithms suitable for FPGA implementation could be exploited [23,24,26–28], possibly leading to lower latency and/or resource occupation. However, MADS proved to be successfully applicable, and comparing different optimization techniques in FPGA is out of the scope of this work. To the authors' knowledge, a digital circuit for real-time control of a power converter that exploits a nonlinear inductor model for prediction is not available in the literature yet. The performances of the proposed circuit are validated through hardware–software co-simulations.

2. Materials and Methods

2.1. Boost Converter Model

We consider the DC–DC boost converter, whose circuit model is shown in Figure 1.

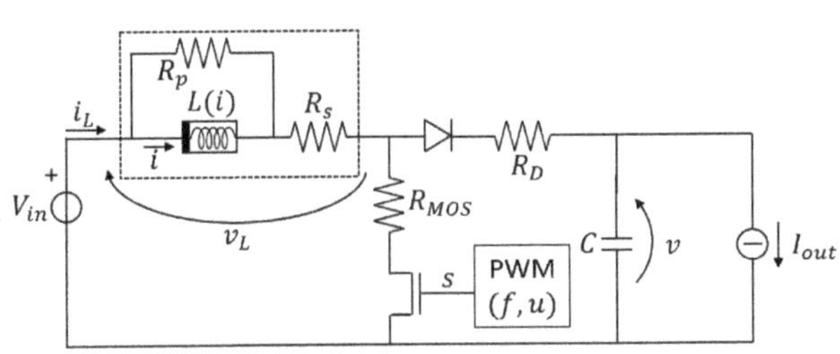

Figure 1. Circuit model of the boost converter.

A pulse width modulation (PWM) signal s with frequency f (period $T = 1/f$) and duty cycle u is applied to the gate of a MOS transistor that behaves like a switch: when $s = 1$ the MOS conducts current (ON phase), through a resistance R_{MOS} that models the conduction losses; when $s = 0$ the transistor is an open circuit and the current flows through the diode, with forward voltage drop v_D and conduction resistance R_D. A load is represented as a variable current source I_{out}, whereas the inductor (enclosed in a dashed rectangle) is modeled through a nonlinear lossless inductor (with differential inductance $L(i)$, flux linkage λ, and current i) and two resistors (R_s and R_p), accounting for all power losses [21]. We are interested in controlling the converter also when the inductor operates in partial saturation, where the differential inductance $L = \frac{d\lambda}{di}$ drops as the current increases:

$$L(i) = L_{sat} + \frac{L_{nom} - L_{sat}}{2}\left\{1 - \frac{2}{\pi}tan^{-1}[\sigma(|i| - I)]\right\}, \tag{1}$$

where parameters R_s, R_p, L_{nom}, L_{sat}, σ, and I are identified starting from experimental measurements of v_L and i_L. Specifically, a subset of these measurements is used for model identification, where the model parameters are determined by solving a nonlinear optimization problem. Another subset of the measurements is used for model validation to assess the accuracy of the identified parameters. All details regarding the measurement process, parameter identification, and validation methodology can be found in [21]. Once the model parameters have been identified, by assuming that the current i can be computed as $i = \Gamma(\lambda)$, the flux linkage λ in the lossless inductor can be evaluated as a function of the current i as

$$\lambda \triangleq \Gamma^{-1}(i) = \int L(i)\frac{di}{dt}dt = \int L(i)di. \tag{2}$$

Even if an analytical expression is available for Γ^{-1}, its evaluation is time- and resource-consuming for an embedded implementation on FPGA. Therefore, 14 couples (i_j, λ_j), $j = 1, \ldots, 14$ are stored in a look-up table (LUT), and functions Γ and Γ^{-1} are computed through linear interpolation.

Because of the fixed-point embedded implementation of the controller, it is convenient to refer to normalized dimensionless quantities. Therefore, we define $\hat{v} = \frac{v}{V_{max}}$, $\hat{V}_{in} = \frac{V_{in}}{V_{max}}$, $\hat{v}_D = \frac{v_D}{V_{max}}$, $\hat{\lambda} = \frac{\lambda}{\Lambda_{max}}$, $\hat{\imath} = \frac{i}{I_{max}} = \frac{\Gamma(\lambda)}{I_{max}} = \frac{\Gamma(\Lambda_{max}\hat{\lambda})}{I_{max}} \triangleq \hat{\Gamma}(\hat{\lambda})$, $\hat{\imath}_L = \frac{i_L}{I_{max}}$, $\hat{I}_{out} = \frac{I_{out}}{I_{max}}$. Coefficients V_{max}, I_{max}, and Λ_{max} are set so that the normalized variables never exceed 1 during the converter operation.

Unlike [21], where i was considered a state variable, here we set the system state as $x = [\hat{\lambda}, \hat{v}]$. The input is u, whereas the vector of measurable parameters is $p = [\hat{V}_{in}, \hat{I}_{out}]$. According to these choices, the continuous-time normalized equations of the boost converter (see Appendix A for the details) are:

$$\frac{dx}{dt} = F(x,u,p) = \begin{cases} \begin{bmatrix} \frac{R_p V_{max}(\hat{V}_{in} - \hat{v}_D - \hat{v}) - (R_s + R_D) R_p I_{max} \hat{\Gamma}(\hat{\lambda})}{(R_s + R_D + R_p)\Lambda_{max}} \\ \frac{V_{max}(\hat{V}_{in} - \hat{v} - \hat{v}_D) + I_{max}[-(R_s + R_D + R_p)\hat{I}_{out} + R_p \hat{\Gamma}(\hat{\lambda})]}{C(R_s + R_D + R_p)V_{max}} \end{bmatrix}, & \text{if } s(u) = 0 \\ \begin{bmatrix} \frac{R_p V_{max} \hat{V}_{in} - R_p (R_s + R_{MOS}) I_{max} \hat{\Gamma}(\hat{\lambda})}{(R_s + R_p + R_{MOS})\Lambda_{max}} \\ \frac{-I_{max}\hat{I}_{out}}{C V_{max}} \end{bmatrix}, & \text{if } s(u) = 1 \end{cases} \tag{3}$$

and the normalized current i_L is

$$\hat{\imath}_L = G(x,u,p) = \begin{cases} \frac{R_p I_{max} \hat{\Gamma}(\hat{\lambda}) + V_{max}(\hat{V}_{in} - \hat{v}_D - \hat{v})}{(R_s + R_D + R_p) I_{max}}, & \text{if } s(u) = 0 \\ \frac{R_p I_{max} \hat{\Gamma}(\hat{\lambda}) + V_{max} V_{in}}{(R_s + R_p + R_{MOS}) I_{max}}, & \text{if } s(u) = 1 \end{cases} \tag{4}$$

Figure 2 shows time plots of the normalized voltage across the lossless inductor (upper panel), $\hat{\lambda}$ (middle panel), and $\hat{\imath}$ (lower panel) at steady state. The curves have been obtained by simulating the boost converter model of Figure 1 with the parameters specified in Section 3 using Simulink and the Simscape Electrical library. The k-th PWM period starts at time $t_k = kT$, when s switches from 0 to 1, and ends at time $t_{k+1} = t_k + T$. The duty cycle $u(t)$ is assumed to be piecewise constant, with value $u(t_k) \triangleq u_k$ in the k-th period ($t_k \leq t < t_{k+1}$). If losses were neglected, the inductor voltage would be a square wave, and the flux linkage would be its integral over time, resulting in a triangular wave. The presence of losses introduces small distortions in both curves. By contrast, current i may be strongly distorted when the inductor operates in partial saturation, due to the nonlinear behavior of the inductance, as shown in the bottom panel.

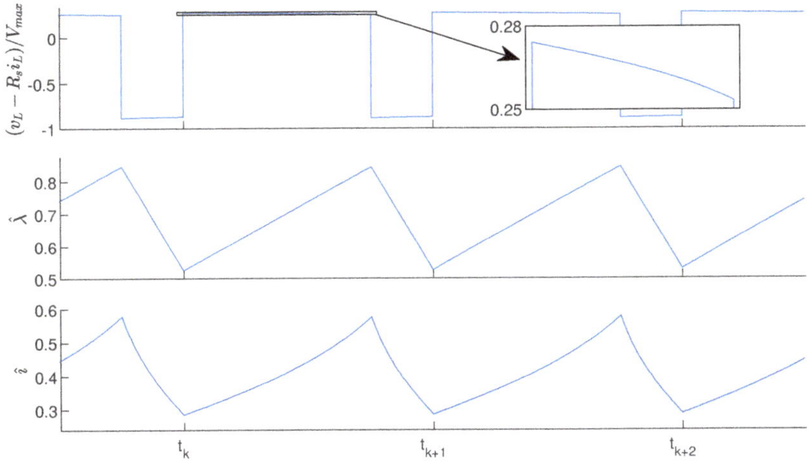

Figure 2. Time plots of the normalized lossless inductor voltage (**top panel**), flux linkage (**middle panel**), and current (**bottom panel**).

2.2. Nonlinear MPC

The aim of the NMPC controller is to keep the average (over a period) output voltage v to a reference value v_{ref}, while satisfying constraints on the duty cycle and on the inductor current [21], namely $u_{low} \leq u \leq u_{high}$ and $i_{low} \leq \hat{\imath}_L \leq i_{high}$. The distance between the normalized reference voltage $\hat{v}_{ref,k}$ and the average (within the k-th PWM period) output voltage $\hat{v}_{avg,k} = \frac{1}{T}\int_{t_k}^{t_{k+1}} \hat{v}(t)dt$ is defined as $\Delta\hat{v}_k = \hat{v}_{avg,k} - \hat{v}_{ref,k}$. We also define $\Delta u_k = u_k - u_{k-1}$.

With NMPC, a nonlinear constrained optimization problem must be solved at each PWM period. The cost function to minimize is computed based on a prediction of the system states over a prediction horizon N. The inputs are optimized up to a control horizon $N_u \leq N$. Many algorithms are available for nonlinear optimization. Among them, zero-order methods (not requiring the evaluation of derivatives) are the most suitable for an embedded implementation. Here we exploit the MADS algorithm, adapted for a digital implementation [25]. This MADS implementation only requires performing operations whose computation is efficient with fixed-point hardware architectures: sums/subtractions, multiplications, shifts, rounding operations, and comparisons. Of course, the computational complexity and latency of the overall optimization algorithm strongly depend on functions F and G (see Equations (3) and (4)), which are evaluated several times at each algorithm iteration.

At time t_k, we use as inputs the measurements of $\hat{v}(t_k)$, $\hat{\imath}_L(t_k)$, $\hat{V}_{in}(t_k)$, $\hat{I}_{out}(t_k)$, the optimal input u_k^* predicted by the MPC at the previous step, and the reference voltage $\hat{v}_{ref,k}$. This means that we assume that p remains constant within the interval NT of the prediction horizon. At the beginning, $u_0^* = u_{low}$. Current $\hat{\imath} = \hat{\Gamma}(\hat{\lambda})$ can be computed by rearranging Equation (4),

$$\hat{\imath} = \begin{cases} \frac{(R_s+R_D+R_p)I_{max}\hat{\imath}_L+V_{max}(-\hat{V}_{in}+\hat{v}_D+\hat{v})}{R_p I_{max}}, & \text{if } s(u)=0 \\ \frac{(R_s+R_p+R_{MOS})I_{max}\hat{\imath}_L-V_{max}V_{in}}{R_p I_{max}}, & \text{if } s(u)=1 \end{cases} \tag{5}$$

Therefore, $\hat{\lambda}(t_k) = \hat{\Gamma}^{-1}(\hat{\imath}(t_k))$. The (N_u-1) optimization variables are gathered in a vector $U = [u_{k+1}, \ldots, u_{k+N_u-1}]$.

The MADS algorithm is run for N_{it} iterations. The reader is referred to [25] and references therein for a detailed explanation of the MADS algorithm. A summary, for ease of reference, is reported in the following. At each iteration, $2(N_u-1)$ poll vectors $U^{(m)}$, $m = 1, \ldots, 2(N_u-1)$, are generated, containing entries within bounds u_{low} and u_{high}. The poll vectors lie on a mesh, inside a frame [31] (see Figure 3). Each poll vector contains the system inputs within the control horizon, i.e., $u_{k+1}, \ldots, u_{k+N_u-1}$. The remaining inputs, up to the prediction horizon, are set as $u_{k+j} = u_{k+N_u-1}$, $j = N_u, \ldots, N-1$. The input sequence can be applied for the integration of system (3), thus obtaining $\hat{\lambda}(t)$, $\hat{v}(t)$, and $\hat{\imath}_L(t)$ (through Equation (4)) up to the prediction horizon. For control purposes, only the values of $\hat{v}$ and $\hat{\imath}_L$ at the PWM switching times are relevant. After computing terms Δu_{k+j} and $\Delta\hat{v}_{k+j}$, the following cost function J can be evaluated, which penalizes both deviations of $\hat{v}$ from its reference value and fast variations of u:

$$J = P\Delta\hat{v}_{k+N}^2 + \sum_{j=1}^{N-1}\left[R\Delta u_{k+j}^2 + Q\Delta\hat{v}_{k+j}^2\right] \tag{6}$$

Since a progressive barrier approach [32] is exploited, a constraint violation function V must also be computed that is equal to 0 when the current i_L is within i_{low} and i_{high}, and grows when the current exceeds the constraints:

$$V = \sum_{t\in\mathcal{T}}\left[\left(\max\left\{0, -\hat{\imath}_L(t) + \frac{i_{low}}{I_{max}}\right\}\right)^2 + \left(\max\left\{0, \hat{\imath}_L(t) - \frac{i_{high}}{I_{max}}\right\}\right)^2\right] \tag{7}$$

Here, $\mathcal{T}$ is the set of all switching instants within the prediction horizon.

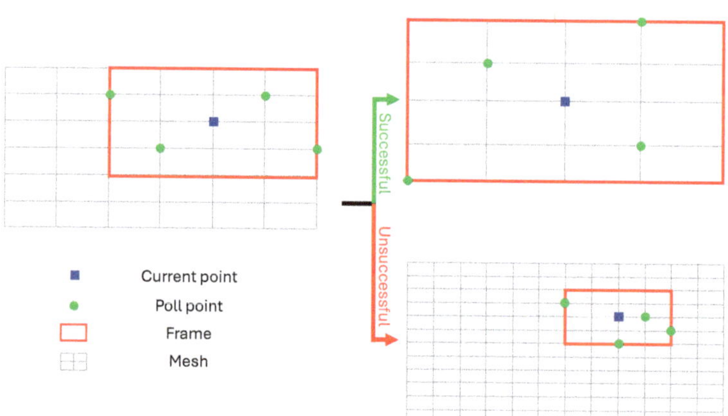

Figure 3. Example of a MADS iteration with $N_u = 3$.

After evaluating all poll points, based on the cost and violation functions, the MADS iteration can be declared successful or not. In the first case, the frame is enlarged and the mesh becomes coarser. In the second case, the opposite happens. A graphical representation (with $N_u = 3$) of a MADS iteration is shown in Figure 3.

At the end of the optimization, after N_{it} iterations, an optimal solution

$$U^* = [u^*_{k+1}, \ldots, u^*_{k+N_u-1}]$$

is obtained. Input u^*_{k+1} is applied to the converter at time t_{k+1}. A new optimization problem is solved through MADS at time t_{k+1} in a receding horizon fashion, by using u^*_{k+1} as a starting input. A timeline of the controller's operation is shown in Figure 4: analog measurements are acquired at time t_k and converted to digital signals with a certain latency (orange rectangle). The optimization problem is then solved (the latency is indicated with the green rectangle) leading to optimal control u^*_{k+1}, which is then applied at time t_{k+1}. This is different from what was done in [21], where the latency was neglected and the output of the optimization at time t_k was u^*_k, applied instantaneously to the boost at time t_k. We remark that the total latency does not affect the control performance, provided that it remains lower than the system sampling time.

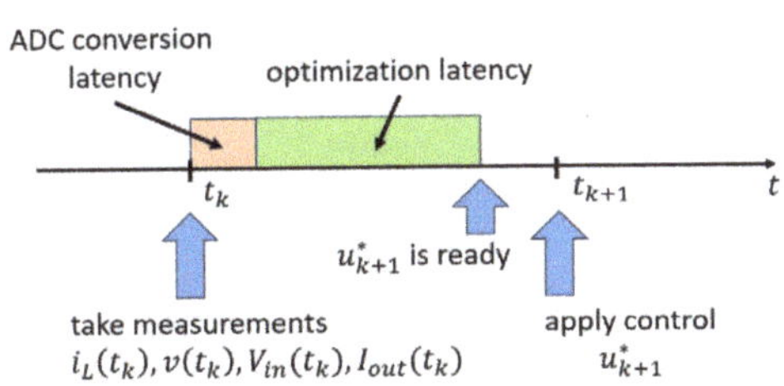

Figure 4. Timeline of the controller's operation.

2.3. Numerical Integration of the Model

In [21], system (3) was solved through the ode45 MATLAB R2023b function, with high accuracy. For a real-time embedded implementation, where the execution time is a major constraint, we have to find an alternative solution.

Consider a generic nonlinear dynamical system $\frac{dx}{dt} = \mathcal{F}(t, x)$. The explicit midpoint method [33] estimates the state at time $t + h$ as

$$x(t+h) = x(t) + h\mathcal{F}\left(t + \frac{h}{2}, x(t) + \frac{1}{2}h\mathcal{F}(t, x(t))\right) \tag{8}$$

This requires the evaluation of function $\mathcal{F}$ at two different points.

We applied the midpoint method to the boost converter with $h = uT$. In our case, the function $\mathcal{F}$ is the function $F(x, u, p)$ defined in Equation (3), under the assumption that both u and p are constant within a period, which is consistent with the discussion presented in Section 2.2. Four function evaluations are necessary within a PWM period. For comparison purposes, we exploit both Equations (3), with $\hat{\lambda}$ as a state variable, and the equations used in [21], where the state variable was $\hat{\imath}$. Figure 5 shows voltage $\hat{v}(t)$ (top panel) and current $\hat{\imath}(t)$ (bottom panel) obtained with ode45 (black curves) and with midpoint methods by exploiting flux linkage $\hat{\lambda}$ and current $\hat{\imath}$ as state variables (see legend). The integration diverges if state $\hat{\imath}$ is used, but good accuracy is obtained with state variable $\hat{\lambda}$. This is because, as shown in Figure 2, the flux linkage is approximately a triangular wave, whereas the current has a cusp-like behavior. Therefore, within two consecutive integration instants, $\frac{d\hat{\lambda}}{dt}$ is approximately constant, unlike $\frac{d\hat{\imath}}{dt}$. This is the reason why we chose $\hat{\lambda}$ as a state variable instead of $\hat{\imath}$ as in [21]. Better performance is obtained using more points, at the cost of a higher computation time. The implementation of the midpoint method for the boost converter is detailed in Appendix B.

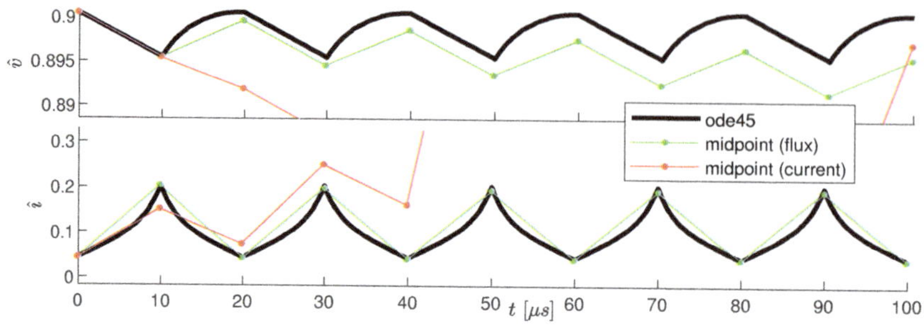

Figure 5. Comparison of ode45 integration, midpoint integration with flux linkage as a state variable, and midpoint integration with current as a state variable.

For solving the optimization problem, the evaluation of the cost function J is the most computationally expensive task, due to numerical integration. Using the midpoint method, the most demanding operation is computing F. The FPGA hardware resources and control algorithm latency strongly depend on the control horizon N, the prediction horizon N_u, and the number of MADS iterations N_{it}. At each iteration, MADS evaluates the cost function at $2(N_u - 1)$ poll points (see Figure 3), and F is computed $4N$ times for each call to the cost function. In summary, at each sampling time, N_F calls to the function F are required, where

$$N_F = 8\, N_{it}\, N\, (N_u - 1). \tag{9}$$

These parameters also affect the control performance, as detailed in the following sections.

2.4. FPGA Implementation

The control algorithm is described in C language through AMD Vitis HLS 2024.2, a high-level synthesis (HLS) tool [34,35], which converts an algorithm coded in C into a fully timed hardware implementation. The workflow consists of the following standard key steps:

1. Compilation;
2. C simulation through a testbench;
3. Register–transfer level (RTL) generation, where the C code is translated into an RTL description, by scheduling operations, binding resources, extracting control logic, and defining external communication;
4. RTL synthesis, which converts the RTL description into a gate-level netlist;
5. RTL simulation through a testbench;
6. Implementation, where the netlist is placed and routed onto device resources, within the logical, physical, and timing constraints.

Directives can be applied to guide the RTL synthesis process starting from the C code. In particular, in most loops we applied pipelining, which is a common practice in digital design to increase the throughput by overlapping sequential arithmetic operations, at the cost of additional resources. Since the J calculation is the algorithm's most computationally demanding part, we unrolled all loops inside the cost function, thus performing arithmetic operations in parallel. This potentially reduces the latency but requires additional resources. Moreover, we used the directive

```
#pragma HLS allocation operation instances = mul limit = Nmul
```

to control the hardware resources. This directive limits the number of multipliers generated in the RTL description. Multipliers are implemented in dedicated digital signal processing slices, which are a limited resource on the FPGA. Therefore, increasing their number can reduce the computation time required by the algorithm, at the cost of using more hardware resources.

A fixed-point data representation is used, through data type <ap_fixed>. Normalized inputs $\hat{\imath}$, $\hat{v}$, $\hat{V}_{in}$, $\hat{I}_{out}$ and output u are represented as unsigned 12-bit numbers with 0 bit of integer part. All internal variables are signed numbers with a variable number of bits for integer and decimal parts, to avoid overflow problems.

We use a Zynq-7000 XC7Z020-1CLG484C FPGA, with a clock frequency of 100 MHz, embedded in a Digilent Zedboard. With the board being equipped with 18-bit multipliers, all multiplications are performed between 18-bit numbers.

3. Results

3.1. Hardware–Software Co-Simulations

The block scheme of the complete system is shown in Figure 6. Measurements of v, i_L, V_{in}, and I_{out} collected on the boost converter are scaled to the voltage range of the analog-to-digital (ADC) converters through, e.g., an analog printed circuit board. The scaling should be such that the maximum voltages V_{max} and current I_{max} map into the maximum ADC voltage value. This way, the digital output of the ADC converters can be interpreted as a fixed-point number with all bits dedicated to the decimal part, leading to the normalized values $\hat{v}, \hat{\imath}_L, \hat{V}_{in}$, and $\hat{I}_{out}$. These signals enter the NMPC block running the MADS algorithm. The reference voltage $\hat{v}_{ref}$ can be provided as a digital input through, e.g., a serial port. The resulting optimal duty cycle value u is provided to a PWM generator and brought to the gate of the MOS transistor (through a proper driver).

In our implementation, only the NMPC block (green) is implemented in the Zynq FPGA. All the other components in Figure 6 are simulated through Simulink R2023b. AMD Vitis Model Composer [36] is exploited to perform this hardware–software co-simulation. Figure 7 shows the adopted setup, with the Zedboard connected to a PC running Model Composer, through a USB cable.

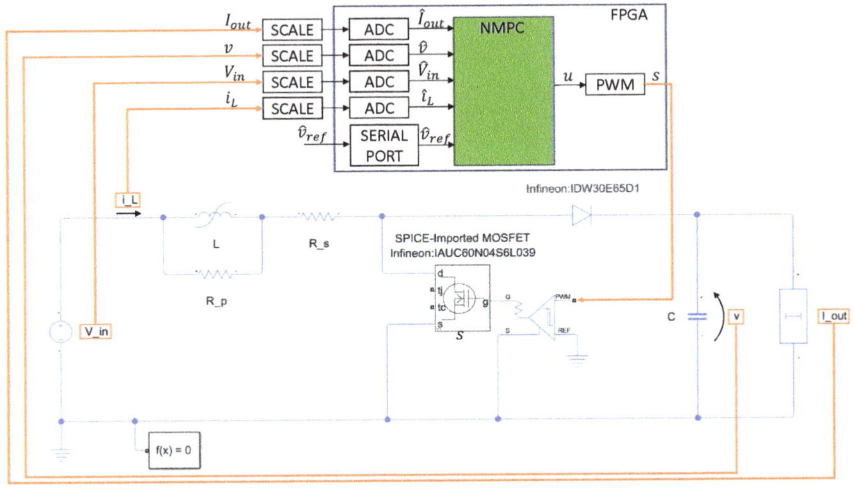

Figure 6. Block scheme of the considered setup.

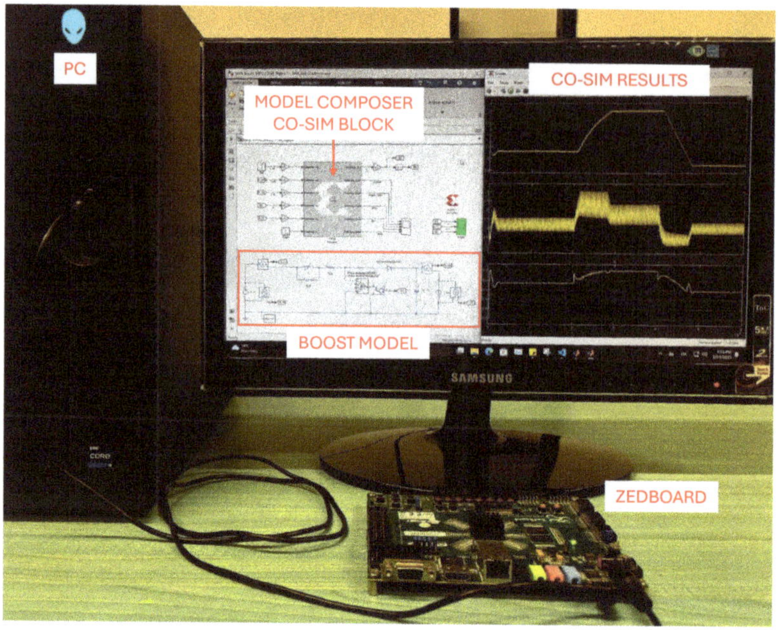

Figure 7. Picture of the hardware–software co-simulation.

The boost converter is modeled using the Simscape Electrical library (see Figure 6). The models of semiconductor devices are based on real components: the Infineon MOSFET IAUC60N04S6L039 [37] and the Infineon diode IDW30E65D1 [38]. A model of the gate driver for the transistor is also included. The MOSFET is modeled using the SPICE netlist provided by the manufacturer, whereas the diode is modeled based on the forward current-voltage curve provided in the datasheet. These models are more accurate than the ones used for MPC prediction, only accounting for conduction resistances R_{MOS} and R_D (see Figure 1).

All considered parameters are listed in Table 1, whereas the piecewise-linear function $\Gamma(\lambda)$ is shown in Figure 8, blue curve. The black dashed curve represents the characteristics of a linear inductor. Notice that, as i approaches the maximum value ($i_{high} = 3$ A), the inductor works in partial saturation and its characteristic drifts apart from the ideal one.

Table 1. System parameters.

Circuit Params.		**NMPC Params.**	
I_{max}	5 A	P	128
V_{max}	6 V	Q	128
Λ_{max}	80 µWb	R	1
C	100 µF	N	5
R_{MOS}	4 mΩ	N_u	2
v_D	0.7 V	u_{low}	0.2
R_D	0.08 Ω	u_{high}	0.8
L_{nom}	35.9848 µH	i_{low}	0 A
L_{sat}	0.5340 µH	i_{high}	3 A
σ_L	1.1704 A^{-1}	**MADS Params.**	
I_L	2.0973 A	N_{it}	7
R_s	0.0462 Ω	**RTL Params.**	
R_p	1.7722 kΩ	N_{mul}	40
f	50 kHz		

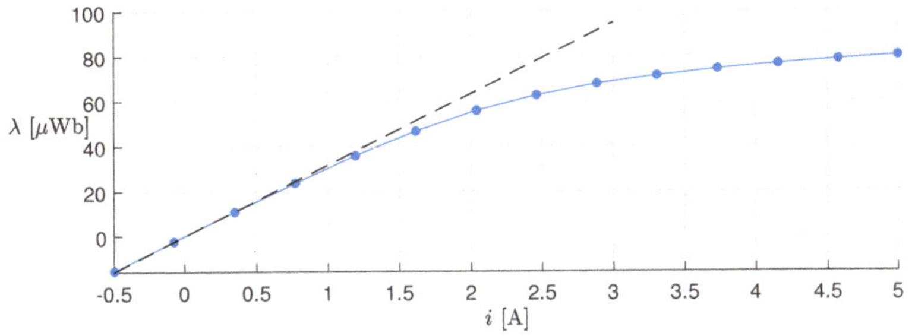

Figure 8. Flux linkage λ vs. current i (blue curve). The dots mark the knee points of the curve. The black dashed line is the ideal (linear) flux linkage–current characteristic.

The values of P, Q, and R, as well as the control horizon N_u and prediction horizon N, are selected through a heuristic process of trial and error, as there is no standard method for determining these values [39]. We remark that P, Q, and R have been chosen as powers of 2 for an efficient hardware implementation. The choice of N_{it} and N_{mul} is discussed later in this section.

The digital circuit performance in terms of latency, used digital signal processors (DSPs), flip flops (FFs), and look-up-tables (LUTs) are listed in Table 2, both after the RTL synthesis and the place and route.

We tested the controller in three different scenarios. In the first test, V_{in} is brought from 0 to 1.8 V in 1 ms (converter startup). Then, it is increased to 2.1 V and decreased again to 1.5 V. The HIL simulation results are shown in Figure 9. The four panels, from top to bottom, show v, i_L, u, and V_{in}, respectively. Notice that v correctly tracks its reference value (black dashed line) and the transients due to the change in V_{in} last about 1.3 ms (see inset). The current and the duty cycle never exceeds the constraints (red dashed lines).

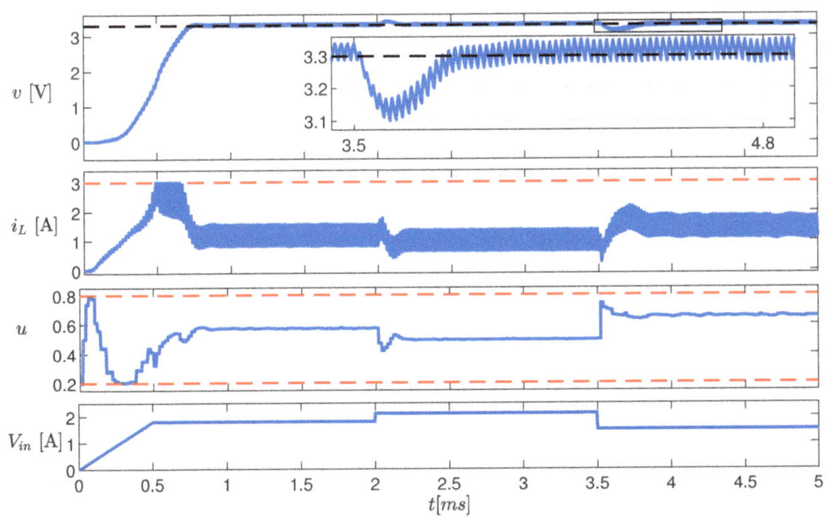

Figure 9. Time evolution of v, i_L, and u (**top three panels**) in response to a variation in V_{in} (**bottom panel**). The red dashed lines are the imposed constraints.

In the second test (Figure 10), I_{out} is changed from 0.5 A to 0.8 A, and than back to 0.4 A. In response to these changes, voltage v exhibits a transient (the first one lasts about 0.5 ms), after which it returns to its reference value. The inductor current hits both the maximum and minimum values, without exceeding them. The constraint $i_L > 0$

prevents the converter from operating in discontinuous conduction mode. The inset shows a cusp-like current waveform, indicating the operation in partial magnetic saturation.

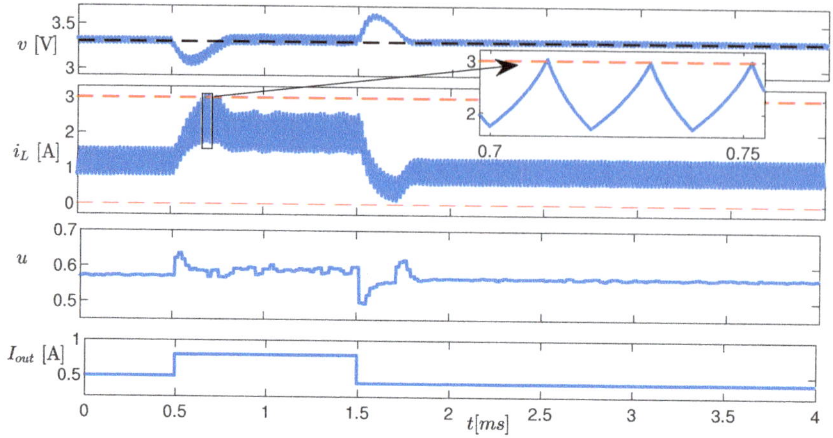

Figure 10. Time evolution of v, i_L, and u (**top three panels**) in response to a variation in I_{out} (**bottom panel**). The red dashed lines are the imposed constraints.

In the last test v_{ref} (black dashed line in the top panel of Figure 11) is changed from 3.3 V to 5 V and then back to 2.7 V. The output voltage is regulated to its reference value in less than 1 ms. The transient time depends on the fact that the current hits the imposed constraints, in both transitions.

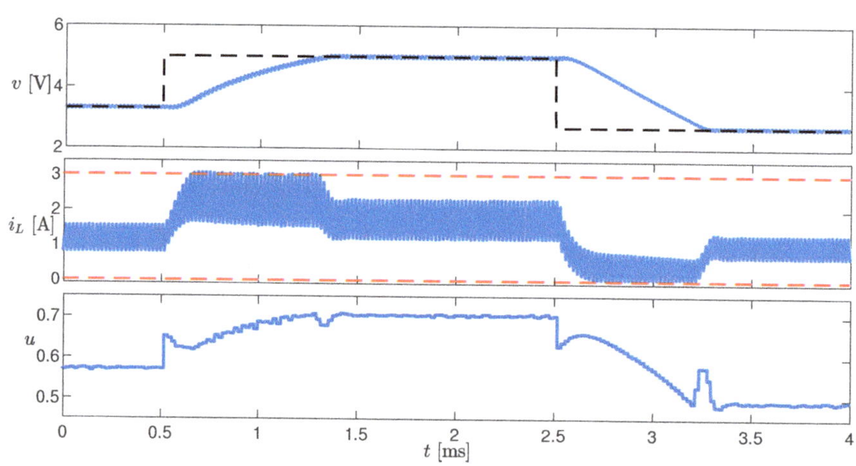

Figure 11. Time evolution of v, i_L, and u in response to a variation in v_{ref} (black dashed line). The red dashed lines are the imposed constraints.

Table 2. Circuit performance.

	Synthesis	Place and Route
latency	16.59 µs	
f_{max}	60.2 kHz	
DSP	53 (24 %)	53 (24 %)
FF	17,387 (16 %)	8791 (8 %)
LUT	25,802 (48 %)	10,655 (20 %)

3.2. Comparisons

The NMPC technique used in this paper was already compared to standard proportional–integral controllers in [21], as well as to MPC where the inductor is modeled as a linear component. Here we compare the performance of the controller proposed in [21], with the one implemented in this paper. We remark that this comparison is not between FPGA implementations. Instead, it shows how the changes made specifically for the FPGA implementation—the MADS optimization algorithm, the application of the control action at the next step (see Figure 4), the simplified numerical integration, and the fixed point representation—affect the control performance. The main differences are listed in Table 3.

Table 3. Main differences between this paper and [21].

	[21]	This Paper
data representation	floating point (64 bit)	fixed point (up to 36 bits)
optimization algorithm	Interior Point (fmincon)	MADS
system integration	ode45	midpoint (ord. 2)
latency	0	16.59 µs
implementation	Simulink	FPGA

Figure 12 shows the simulation results in response to a change in v_{ref} obtained in [21] as a benchmark, and the new results obtained with the hardware–software co-simulation, with a different number N_{it} of iterations of the MADS algorithm (see legend). The delay with respect to the benchmark case is mainly due to the fact that the control obtained based on measurements at time t_k is applied at time t_{k+1}. Therefore, the control response to a change in v_{ref} is delayed with respect to sampling time. If N_{it} decreases, a suboptimal control is applied, resulting in a larger delay. With $N_{it} = 2$, the stationary steady state is not reached, whereas for $N_{it} > 7$ there is no significant improvement. With $N_{it} = 7$, the delay in v to reach the setpoint is about 100 µs.

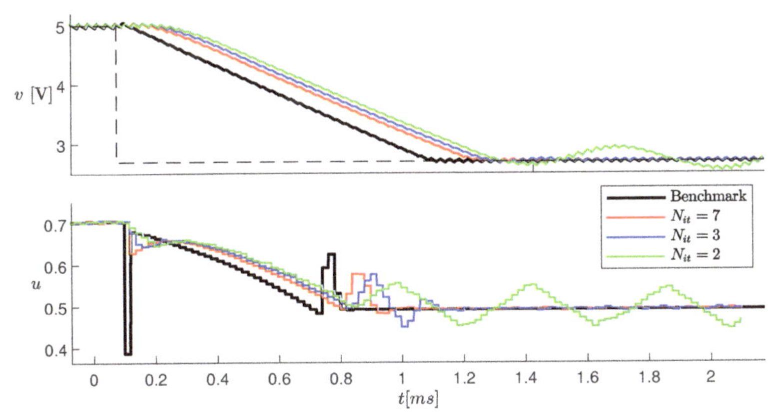

Figure 12. Simulation results with different values of N_{it}. The black curves are related to the benchmark case.

Increasing N_{it}, on the other hand, impacts the circuit latency, as detailed in the next section.

3.3. Circuit Performance

In this section, we show the digital circuit performance in response to changes in some parameters. Figure 13 shows, from top to bottom, the latency of the control algorithm, the percentage of used DSPs, FFs, and LUTs versus value N_{mul} applied to the `#pragma` directive (see Section 2.4). The latency decreases and the resource occupation increases with N_{mul}, as expected. However, for $N_{mul} > 40$, the resources continue to grow, but the latency remains constant to about 16.5 µs. For this reason, we chose $N_{mul} = 40$ in our implementation.

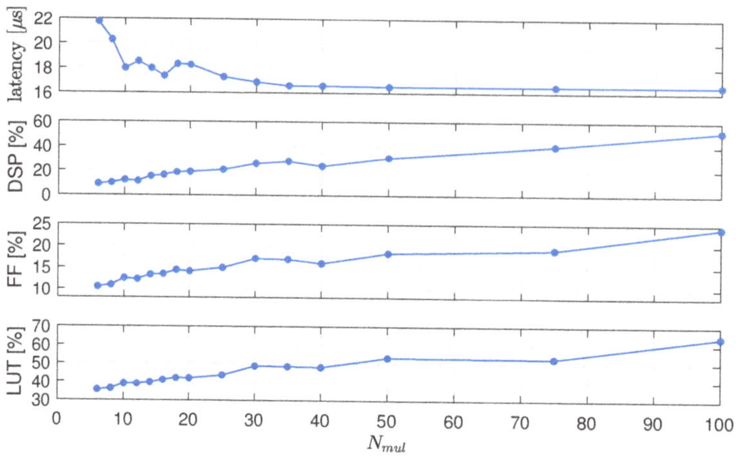

Figure 13. From top to bottom: latency of the control algorithm, percentage of used DSPs, FFs, and LUTs vs. N_{mul}.

With $N_{mul} = 40$, the RTL synthesis was performed for several combinations of N, N_u, and N_{it}. For each of them, the latency and the FPGA resource occupation are listed in Table 4. The bold line refers to the parameters used in Section 3.1. Latency increases linearly with N_{it}, while resource occupation remains roughly constant. By increasing N and N_u, the latency also increases, and the resource occupation tends to grow as well, especially LUT usage, which reaches 100% in the Zynq FPGA for $N = 15$. This is due to the number N_F of evaluations of the F function, which is directly proportional to N_{it}, N, and N_u (see Equation (9)). We remark that not all parameters' combinations listed in the table lead to good control performances, as shown in the next section. A unified design of both the control algorithm and the digital circuit is then necessary to meet all specifications.

Table 4. Latency and FPGA resource utilization for different values of N, N_u, and N_{it}. The bold line refers to the parameters used in Section 3.1.

N	N_u	N_{it}	Latency (µs)	DSP (%)	FF (%)	LUT (%)
4	2	3	6.00	68 (31%)	18,203 (17%)	24,252 (48%)
4	2	5	9.86	68 (31%)	18,205 (17%)	24,254 (48%)
4	2	7	13.72	68 (31%)	18,205 (17%)	24,254 (48%)
4	3	3	7.52	65 (29%)	19,349 (18%)	25,991 (48%)
4	3	5	12.28	65 (29%)	19,351 (18%)	25,993 (48%)
4	3	7	17.04	65 (29%)	19,351 (18%)	25,993 (48%)
5	2	3	7.23	53 (24%)	17,389 (16%)	25,800 (48%)
5	2	5	11.91	53 (24%)	17,391 (16%)	25,802 (48%)
5	**2**	**7**	**16.59**	**53 (24%)**	**17,837 (16%)**	**25,802 (48%)**
5	3	3	8.72	72 (32%)	21,639 (20%)	29,090 (54%)
5	3	5	14.28	72 (32%)	21,641 (20%)	29,092 (54%)
5	3	7	19.84	72 (32%)	21,641 (20%)	29,092 (54%)

Table 4. *Cont.*

N	N_u	N_{it}	Latency (µs)	DSP (%)	FF (%)	LUT (%)
6	2	3	8.43	65 (29%)	20,765 (19%)	29,981 (56%)
6	2	5	13.91	65 (29%)	20,767 (19%)	29,983 (56%)
6	2	7	19.39	65 (29%)	20,767 (19%)	29,983 (56%)
6	3	3	10.04	74 (33%)	24,192 (22%)	33,609 (63%)
6	3	5	16.48	74 (33%)	24,194 (22%)	33,611 (63%)
6	3	7	22.92	74 (33%)	24,194 (22%)	33,611 (63%)
7	2	3	9.66	79 (35%)	24,316 (22%)	34,597 (65%)
7	2	5	15.96	79 (35%)	23,418 (22%)	34,599 (65%)
7	2	7	22.26	79 (35%)	23,418 (22%)	34,599 (65%)
7	3	3	11.33	64 (29%)	23,569 (22%)	34,801 (65%)
7	3	5	18.63	64 (29%)	23,571 (22%)	34,803 (65%)
7	3	7	25.93	64 (29%)	23,571 (22%)	34,803 (65%)
10	2	3	13.41	73 (33%)	27,179 (25%)	41,218 (77%)
10	2	5	22.21	73 (33%)	27,181 (25%)	41,220 (77%)
10	2	7	31.01	73 (33%)	27,181 (25%)	41,220 (77%)
10	3	3	15.02	75 (34%)	28,058 (26%)	42,719 (80%)
10	3	5	25.08	75 (34%)	28,058 (26%)	42,719 (80%)
10	3	7	34.96	75 (34%)	28,060 (26%)	42,721 (80%)
13	2	3	17.10	75 (34%)	30,191 (28%)	49,552 (93%)
13	2	5	28.36	75 (34%)	30,193 (28%)	49,554 (93%)
13	2	7	39.62	75 (34%)	30,193 (28%)	49,554 (93%)
13	3	3	19.13	71 (32%)	31,090 (29%)	50,886 (95%)
13	3	5	31.63	71 (32%)	31,092 (29%)	50,888 (95%)
13	3	7	44.13	71 (32%)	31,092 (29%)	50,888 (95%)
15	2	3	19.44	72 (32%)	31,770 (29%)	53,657 (100%)
15	2	5	32.36	72 (32%)	31,772 (29%)	53,659 (100%)
15	2	7	45.08	72 (32%)	31,774 (29%)	53,661 (100%)
15	3	3	21.83	78 (35%)	34,931 (32%)	57,180 (107%)
15	3	5	36.13	78 (35%)	34,933 (32%)	57,182 (107%)
15	3	7	50.43	78 (35%)	34,933 (32%)	57,182 (107%)

The effect on control when varying N_{it} has already been discussed in Section 3.2. Here, the controller is tested by varying N and N_u. Figure 14 shows some simulation results when $N_{it} = 5$, $N_u = 2$, and v_{ref} changes as shown in Section 3.1, for both $N = 4$ and $N = 5$. When $N = 4$, the prediction horizon is too short, and the controller is ineffective; i.e., v does not reach v_{ref}. If N is sufficiently large—in particular, for $N > 4$—the controller is effective. It has been verified that increasing N beyond 5 does not further improve controller performance. Similarly, increasing N_u from 2 to 3 does not impact control performance.

Some latencies in Table 4 are sufficiently low to allow an increase in the converter switching frequency f over 50 kHz. For example, let us consider the case with $N = 7$, $N_u = 2$, and $N_{it} = 3$, corresponding to a latency of 9.66 µs. In Figure 15, the scenario where v_{ref} changes is shown for converters operating at switching frequencies of 50 kHz, 60 kHz, 80 kHz, and 100 kHz. Up to 80 kHz, the controller behavior is almost unchanged, with only a slight difference in the settling time of the converter output voltage (see the inset). At 100 kHz, the controller is ineffective. In all simulations, the prediction horizon is $N = 7$, whereas the sampling time changes with the frequency. With $f = 50$ kHz, the controller makes a prediction for the next $N/f = 140$ µs; with $f = 100$ kHz, this interval becomes shorter ($N/f = 70$ µs). Although the algorithm latency would allow operation up to $f = 100\,\text{kHz}$, the prediction horizon N is not long enough to ensure acceptable control performance.

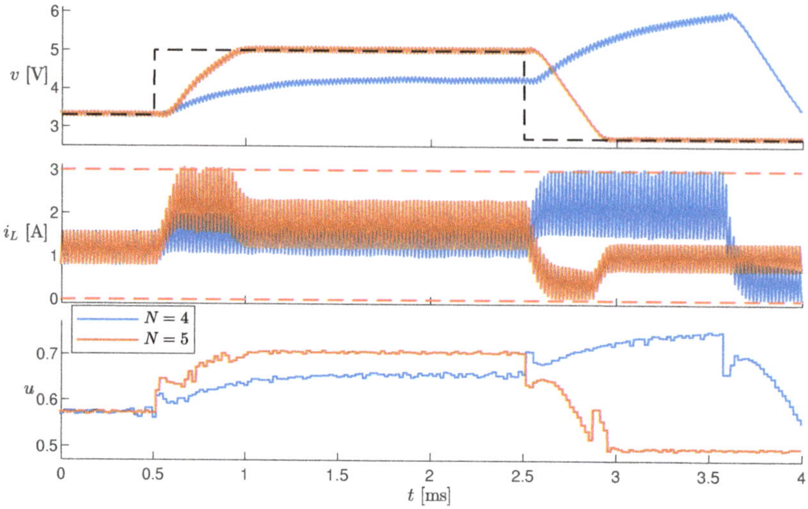

Figure 14. Simulation results with different prediction horizons N.

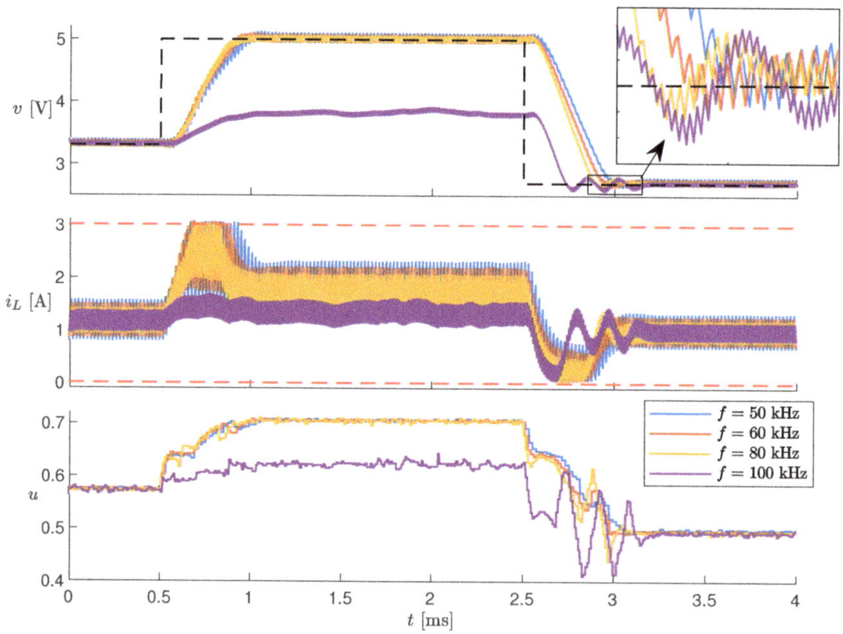

Figure 15. Simulation results with different switching frequencies f.

4. Discussion

Traditionally, in power converters, inductor saturation is avoided by assuming a constant inductance, enabling straightforward predictions of power losses and current ripple. Under these conditions, both PI (model-free) regulators and model-based controllers, such as MPC, are effective for converter control. When the model is sufficiently accurate, MPC typically outperforms PI controllers by inherently enforcing state and input constraints.

To enhance power density in power converters, smaller inductors and higher switching frequencies can be employed to reduce current ripple and prevent saturation. In such cases, MPC requires an accurate inductor model to predict behavior near saturation. The technique proposed in [21] demonstrates that the nonlinear behavioral inductor model [9] can be effectively integrated into NMPC for voltage regulation in switching converters, even when inductors operate in partial saturation, by enforcing constraints and outperforming standard PI regulators. Conversely, using a conventional inductor model with constant inductance leads to constraint violations. This work makes a step forward with respect to [21], by implementing the NMPC controller on an FPGA and testing it through hardware–software co-simulations. This allows including the effect of data quantization, fixed-point representation, and latency, as well as the possibility to exploit simpler integration methods (e.g., the explicit midpoint) and derivative-free optimization algorithms (e.g., MADS).

This paper provides a proof of concept about the possibility of applying NMPC up to PWM frequencies of about 80 kHz. It should be noted that the presented results are valid for the specific converter used in this work. When employing a different converter, its dynamics may vary, so the maximum achievable switching frequency may vary. Therefore, the selection of the parameters N, N_u, and N_{it} must be evaluated case by case.

The next step will be to apply the embedded controller to a real boost converter.

5. Conclusions

In this work, an NMPC technique for the control of a boost converter with a nonlinear inductor is implemented on an FPGA. The hardware–software co-simulation results show that the embedded controller is able to regulate the converter's output voltage by fulfilling current constraints, even when the inductor operates at partial magnetic saturation, up to PWM frequencies of about 80 kHz. Further work will be concerned with the real-time control of a real boost converter, thus assessing its robustness against measurement noise and model inaccuracies.

Author Contributions: Conceptualization, A.O.; Methodology, M.L.; Software, A.R. and A.O.; Validation, A.R.; Data curation, M.L.; Writing—original draft, A.O.; Writing—review & editing, M.S.; Supervision, M.S. All authors have read and agreed to the published version of the manuscript.

Funding: This work was partially funded by the European Union-NextGenerationEU, within the project "MAGSAT-Exploiting MAGnetic SATuration to increase power density in switching converters", University of Genoa, Italy.

Data Availability Statement: Data are contained within the article.

Conflicts of Interest: The authors declare no conflicts of interest.

Appendix A

We start from Equations (5) and (6) in [21], where the state variables are v and i. Here, we perform a change of variables by considering that $i = \Gamma(\lambda)$ and $L(i)\frac{di}{dt} = \frac{d\lambda}{dt}$. Therefore, we obtain the differential equations for state $x = \begin{bmatrix} \lambda \\ v \end{bmatrix}$:

$$\begin{bmatrix} \frac{d\lambda}{dt} \\ \frac{dv}{dt} \end{bmatrix} = \begin{cases} \begin{bmatrix} \frac{R_p(V_{in}-v_D-v)-(R_s+R_D)R_p\Gamma(\lambda)}{(R_s+R_D+R_p)} \\ \frac{(V_{in}-v-v_D)+[-(R_s+R_D+R_p)I_{out}+R_p\Gamma(\lambda)]}{C(R_s+R_D+R_p)} \end{bmatrix}, & \text{if } s(u)=0 \\ \begin{bmatrix} \frac{R_pV_{in}-R_p(R_s+R_{MOS})\Gamma(\lambda)}{(R_s+R_p+R_{MOS})} \\ \frac{-I_{out}}{C} \end{bmatrix}, & \text{if } s(u)=1 \end{cases} \tag{A1}$$

The current i_L can be expressed as

$$i_L = \begin{cases} \frac{R_p\Gamma(\lambda)+(V_{in}-v_D-v)}{(R_s+R_D+R_p)}, & \text{if } s(u)=0 \\ \frac{R_p\Gamma(\lambda)+V_{in}}{(R_s+R_p+R_{MOS})}, & \text{if } s(u)=1 \end{cases} \tag{A2}$$

In addition to [21], we include the diode resistance R_D. Now, we refer to normalized variables $\hat{v} = \frac{v}{V_{max}}$, $\hat{V}_{in} = \frac{V_{in}}{V_{max}}$, $\hat{v}_D = \frac{v_D}{V_{max}}$, $\hat{\lambda} = \frac{\lambda}{\Lambda_{max}}$, $\hat{\imath} = \frac{i}{I_{max}} = \frac{\Gamma(\lambda)}{I_{max}} = \frac{\Gamma(\Lambda_{max}\hat{\lambda})}{I_{max}} \triangleq \hat{\Gamma}(\hat{\lambda})$, $\hat{\imath}_L = \frac{i_L}{I_{max}}$, $\hat{I}_{out} = \frac{I_{out}}{I_{max}}$. By substituting these terms in Equations (A1) and (A2), we obtain Equations (3) and (4).

Appendix B

For the FPGA implementation, it is important to spare time and resources. Therefore, we define the following (dimensionless) constants that can be computed offline and stored in the circuit memory.

$$\alpha = \frac{R_p}{R_s + R_{MOS} + R_p}, \ \beta = \frac{R_p}{R_s + R_D + R_p} \tag{A3}$$

$$C_1 = -\frac{\alpha(R_s + R_{MOS})I_{max}}{\Lambda_{max}}T, \ C_2 = \frac{\alpha V_{max}}{\Lambda_{max}}T, \ C_3 = -\frac{I_{max}}{CV_{max}}T, \tag{A4}$$

$$C_4 = -\frac{\beta(R_s + R_D)I_{max}}{\Lambda_{max}}T, \ C_5 = -\frac{\beta V_{max}}{\Lambda_{max}}T, \ C_6 = -\frac{\beta V_{max}\hat{v}_D}{\Lambda_{max}}T, \tag{A5}$$

$$C_7 = \frac{\beta I_{max}}{CV_{max}}T, \ C_8 = -\frac{\beta}{CR_p}T, \ C_9 = -\frac{I_{max}}{CV_{max}}T, \ C_{10} = -\frac{\beta\hat{v}_D}{CR_p}T \tag{A6}$$

The parameters $C_1, \ldots, C_{10}$ depend on T; therefore, they must be updated if the converter's switching frequency is changed.

Recall that, in the MPC prediction phase, the values of V_{in} and I_{out} are assumed to be constant. By referring to Equation (3), the integration is performed through the following operations.

$$K_1 = \begin{bmatrix} K_{11} \\ K_{12} \end{bmatrix} = u_k \begin{bmatrix} C_1\Gamma[\hat{\lambda}(t_k)] + C_2\hat{V}_{in}(t_k) \\ C_3\hat{\imath}_{out}(t_k) \end{bmatrix} \tag{A7}$$

$$K_2 = u_k \begin{bmatrix} C_1\Gamma[\hat{\lambda}(t_k) + \frac{1}{2}K_{11}] + C_2\hat{V}_{in}(t_k) \\ C_3\hat{\imath}_{out}(t_k) \end{bmatrix} \tag{A8}$$

$$t_k' = t_k + u_k T, \quad x(t_k') = \begin{bmatrix} \lambda(t_k') \\ v(t_k') \end{bmatrix} = x(t_k) + K_2 \tag{A9}$$

$$K_1 = \begin{bmatrix} K_{11} \\ K_{12} \end{bmatrix} = (1-u_k) \begin{bmatrix} C_4\Gamma[\hat{\lambda}_k'] + C_5[\hat{v}(t_k) + \hat{V}_{in}(t_k)] + C_6 \\ C_7\Gamma[\hat{\lambda}(t_k')] + C_8[\hat{v}(t_k) + \hat{V}_{in}(t_k)] + C_9\hat{V}_{in}(t_k) + C_{10} \end{bmatrix} \tag{A10}$$

$$K_2 = (1-u_k) \begin{bmatrix} C_4\Gamma[\hat{\lambda}(t_k') + \frac{1}{2}K_{11}] + C_5[\hat{v}(t_k) + \hat{V}_{in}](t_k) + C_6 \\ C_7\Gamma[\hat{\lambda}(t_k') + \frac{1}{2}K_{11}] + C_8[\hat{v}(t_k) + \hat{V}_{in}(t_k)] + C_9\hat{\imath}_{out}(t_k) + C_{10} \end{bmatrix} \tag{A11}$$

$$x(t_{k+1}) = \begin{bmatrix} \lambda(t_{k+1}) \\ v(t_{k+1}) \end{bmatrix} = x(t_k') + K_2 \tag{A12}$$

References

1. Milner, L.; Rincón-Mora, G.A. Small saturating inductors for more compact switching power supplies. *IEEJ Trans. Electr. Electron. Eng.* **2012**, *7*, 69–73. [CrossRef]
2. Di Capua, G.; Femia, N. A novel method to predict the real operation of ferrite inductors with moderate saturation in switching power supply applications. *IEEE Trans. Power Electron.* **2016**, *31*, 2456–2464. [CrossRef]

3. Di Capua, G.; Femia, N.; Stoyka, K. Impact of Inductors Saturation on DC–DC Switching Regulators. In Proceedings of the IEEE International Forum on Research and Technology for Society and Industry (RTSI), Florence, Italy, 9–12 September 2019; pp. 254–259.
4. Scirè, D.; Lullo, G.; Vitale, G. Assessment of the Current for a Non-Linear Power Inductor Including Temperature in DC–DC Converters. *Electronics* **2023**, *12*, 579. [CrossRef]
5. Oliveri, A.; Lodi, M.; Storace, M. Nonlinear models of power inductors: A survey. *Int. J. Circuit Theory Appl.* **2022**, *1*, 2–34. [CrossRef]
6. Scirè, D.; Lullo, G.; Vitale, G. Non-Linear Inductor Models Comparison for Switched-Mode Power Supplies Applications. *Electronics* **2022**, *11*, 2472. [CrossRef]
7. Kaiser, J.; Duerbaum, T. An overview of saturable inductors: Applications to power supplies. *IEEE Trans. Power Electron.* **2021**, *36*, 10766–10775. [CrossRef]
8. Ravera, A.; Formentini, A.; Lodi, M.; Oliveri, A.; Passalacqua, M.; Storace, M. Modeling the Effect of Air-Gap Length and Number of Turns on Ferrite-Core Inductors Working up to Magnetic Saturation in a Buck Converter. *IEEE Trans. Circ. Syst. I* **2024**, *71*, 5400–5409. . [CrossRef]
9. Ravera, A.; Oliveri, A.; Lodi, M.; Storace, M. A nonlinear behavioral model of a ferrite-core inductor with fixed-frequency sinusoidal voltage input. In Proceedings of the IEEE International Conference on Smart Techonologies (EUROCON), Turin, Italy, 6–8 July 2023.
10. Ravera, A.; Oliveri, A.; Lodi, M.; Beatrice, C.; Ferrara, E.; Fiorillo, F.; Storace, M. Modeling Amorphous-Core Inductors up to Magnetic Saturation. *IEEE Trans. Power Electron.* **2024**, *40*, 1563–1576. [CrossRef]
11. Ravera, A.; Oliveri, A.; Lodi, M.; Bemporad, A.; Heemels, W.; Kerrigan, E.C.; Storace, M. Co-Design of a Controller and Its Digital Implementation: The MOBY-DIC2 Toolbox for Embedded Model Predictive Control. *IEEE Trans. Control Syst. Technol.* **2023**, *31*, 2871–2878. [CrossRef]
12. Karamanakos, P.; Geyer, T.; Oikonomou, N.; Kieferndorf, F.D.; Manias, S. Direct model predictive control: A review of strategies that achieve long prediction intervals for power electronics. *IEEE Ind. Electron. Mag.* **2014**, *8*, 32–43. [CrossRef]
13. Muthukumar, P.; Venkateshkumar, M.; Chin, C.S. A Comparative Study of Cutting-Edge Bi-Directional Power Converters and Intelligent Control Methodologies for Advanced Electric Mobility. *IEEE Access* **2024**, *12*, 28710–28752. [CrossRef]
14. Liu, Y.C.; Ge, X.; Tang, Q.; Deng, Z.; Gou, B. Model predictive current control for four-switch three-phase rectifiers in balanced grids. *Electron. Lett.* **2017**, *53*, 44–46. [CrossRef]
15. Jin, T.; Wei, H.; Legrand, D.; Nzongo, M.; Zhang, Y. Model predictive control strategy for NPC grid-connected inverters in unbalanced grids. *Electron. Lett.* **2016**, *52*, 1248–1250. [CrossRef]
16. Wu, J.; Li, J.; Wang, H.; Li, G.; Ru, Y. Fault-Tolerant Three-Vector Model-Predictive-Control-Based Grid-Connected Control Strategy for Offshore Wind Farms. *Electronics* **2024**, *13*, 2316. [CrossRef]
17. Bououden, S.; Hazil, O.; Filali, S.; Chadli, M. Modelling and model predictive control of a DC–DC Boost converter. In Proceedings of the 2014 15th International Conference on Sciences and Techniques of Automatic Control and Computer Engineering (STA), Hammamet, Tunisia, 21–23 December 2024; IEEE: Piscataway, NJ, USA, 2014; pp. 643–648.
18. Lekić, A.; Hermans, B.; Jovičić, N.; Patrinos, P. Microsecond nonlinear model predictive control for DC–DC converters. *Int. J. Circuit Theory Appl.* **2020**, *48*, 406–419. [CrossRef]
19. Stickan, B.; Frison, G.; Burger, B.; Diehl, M. A nonlinear Real-Time Pulse-Pattern MPC Scheme for Power-Electronics Circuits Operating in the Microseconds Range. In Proceedings of the American Control Conference, Atlanta, GA, USA, 8–10 June 2022; pp. 2070–2077.
20. Mariethoz, S.; Herceg, M.; Kvasnica, M. Model Predictive Control of buck DC–DC converter with nonlinear inductor. In Proceedings of the 11th Workshop on Control and Modeling for Power Electronics, Zurich, Switzerland, 17–20 August 2008; pp. 1–8.
21. Firpo, P.; Ravera, A.; Oliveri, A.; Lodi, M.; Storace, M. Use of a Partially Saturating Inductor in a Boost Converter with Model Predictive Control. *Electronics* **2023**, *12*, 3013. [CrossRef]
22. McInerney, I.; Constantinides, G.A.; Kerrigan, E.C. A Survey of the Implementation of Linear Model Predictive Control on FPGAs**The support of the EPSRC Centre for Doctoral Training in High Performance Embedded and Distributed Systems (HiPEDS, Grant Reference EP/L016796/1) is gratefully acknowledged. *IFAC-PapersOnLine* **2018**, *51*, 381–387. [CrossRef]
23. Xu, F.; Guo, Z.; Chen, H.; Ji, D.; Qu, T. A Custom Parallel Hardware Architecture of Nonlinear Model Predictive Control on FPGA. *IEEE Trans. Ind. Electron.* **2022**, *69*, 11569–11579. [CrossRef]
24. Guo, H.; Liu, F.; Xu, F.; Chen, H.; Cao, D.; Ji, Y. Nonlinear model predictive lateral stability control of active chassis for intelligent vehicles and its FPGA implementation. *IEEE Trans. Syst. Man, Cybern. Syst.* **2017**, *49*, 2–13. [CrossRef]
25. Ravera, A.; Oliveri, A.; Lodi, M.; Storace, M. MADS-based fast FPGA implementation of nonlinear model predictive control. In Proceedings of the IEEE International Symposium on Circuits and Systems, Monterey, CA, USA, 21–25 May 2023.

26. Englert, T.; Völz, A.; Mesmer, F.; Rhein, S.; Graichen, K. A software framework for embedded nonlinear model predictive control using a gradient-based augmented Lagrangian approach (GRAMPC). *Optim. Eng.* **2019**, *20*, 769–809. [CrossRef]
27. Hausberger, T.; Kugi, A.; Eder, A.; Kemmetmüller, W. High-speed nonlinear model predictive control of an interleaved switching DC/DC-converter. *Control Eng. Pract.* **2020**, *103*, 104576. [CrossRef]
28. Patne, V.; Ingole, D.; Sonawane, D. FPGA Implementation Framework for Low Latency Nonlinear Model Predictive Control. *IFAC-PapersOnLine* **2020**, *53*, 7020–7025. [CrossRef]
29. Ravera, A.; Oliveri, A.; Lodi, M.; Storace, M. Embedded Linear Model Predictive Control Through Mesh Adaptive Direct Search Algorithm. In Proceedings of the 2019 26th IEEE International Conference on Electronics, Circuits and Systems (ICECS), Genoa, Italy, 27–29 November 2019; pp. 542–545. [CrossRef]
30. Khusainov, B.; Kerrigan, E.C.; Constantinides, G.A. Automatic Software and Computing Hardware Codesign for Predictive Control. *IEEE Trans. Control Syst. Technol.* **2019**, *27*, 2295–2304. [CrossRef]
31. Audet, C.; Dennis, J.E., Jr. Mesh adaptive direct search algorithms for constrained optimization. *SIAM J. Optim.* **2006**, *17*, 188–217. [CrossRef]
32. Audet, C.; Dennis, J.E. A Progressive Barrier for Derivative-Free Nonlinear Programming. *SIAM J. Optim.* **2009**, *20*, 445–472. [CrossRef]
33. Süli, E.; Mayers, D.F. *An Introduction to Numerical Analysis*; Cambridge University Press: Cambridge, UK, 2003.
34. AMD. *Vitis Reference Guide*; Technical Report UG1702 (v2024.2); AMD, Inc.: Sunnyvale, CA, USA, 2025.
35. Lucia, S.; Navarro, D.; Lucia, O.; Zometa, P.; Findeisen, R. Optimized FPGA implementation of model predictive control for embedded systems using high-level synthesis tool. *IEEE Trans. Ind. Inform.* **2017**, *14*, 137–145. [CrossRef]
36. AMD. *Vitis Model Composer Reference Guide*; Technical Report UG1483 (v2024.2); AMD, Inc.: Sunnyvale, CA, USA, 2025.
37. Infineon. OptiMOS™-6 Power-Transistor—IAUC60N04S6L039. 2019. Available online: https://www.infineon.com/dgdl/Infineon-IAUC60N04S6L039-DataSheet-v01_00-EN.pdf?fileId=5546d4626afcd350016b1c511cb70df6 (accessed on 24 February 2025).
38. Infineon. IDW30E65D1—Emitter Controlled Diode Rapid 1 Series. 2013. Available online: https://www.infineon.com/dgdl/Infineon-IDW30E65D1-DS-v02_02-en.pdf?fileId=db3a30433d68e984013d6929d1da01cc (accessed on 24 February 2025).
39. Grüne, L.; Pannek, J. *Nonlinear Model Predictive Control*; Springer: Berlin/Heidelberg, Germany, 2017.

electronics

MDPI

Article

Dual-Random Space Vector Pulse Width Modulation Strategy Based on Optimized Beta Distribution

Xin Gu [1], Kunyang Wu [1], Xuefeng Jin [1,*], Guozheng Zhang [1], Wei Chen [1] and Chen Li [2]

1 School of Electrical Engineering, Tiangong University, Tianjin 300387, China; guxin@tiangong.edu.cn (X.G.); wukunyang@tiangong.edu.cn (K.W.); zhanggz@tju.edu.cn (G.Z.); chen_wei@tiangong.edu.cn (W.C.)

2 Advanced Electrical Equipment Innovation Center, Zhejiang University, Hangzhou 311107, China; lichen_hz@zju.edu.cn

* Correspondence: jinxuefeng@tiangong.edu.cn

Abstract: In the control system of a permanent magnet synchronous motor (PMSM) driven by an inverter, the conventional space vector pulse width modulation (SVPWM) strategy introduces high-frequency current harmonics at the switching frequency and its multiples, resulting in significant high-frequency vibrations during motor operation. To address this issue, a dual-random SVPWM strategy is proposed in this paper, which combines a random switching frequency and random zero-vector to spread the spectrum of high-frequency current harmonics. This approach effectively disperses the high-frequency harmonics concentrated at the switching frequency and its multiples, thereby significantly reducing the motor's high-frequency vibrations. Furthermore, to overcome the limitations of the traditional linear congruential method in generating random numbers, the Beta distribution is introduced and improved in this study. The particle swarm optimization (PSO) algorithm is employed to optimize the shape parameters of the Beta distribution, to achieve the optimal random number performance. Finally, experimental validation is conducted under various speed conditions. Compared with the conventional SVPWM strategy, the results demonstrate that the proposed dual-random SVPWM strategy exhibits superior suppression of both high-frequency harmonics and high-frequency vibrations.

Keywords: permanent magnet synchronous motor; current harmonics; dual-random SVPWM; particle swarm optimization; high-frequency vibration

1. Introduction

Permanent magnet synchronous motors (PMSMs) are increasingly widely used in fields such as new energy vehicles (EVs), industrial automation, and rail transit, owing to their advantages of high-power density, high efficiency, and low maintenance costs. With the continuous expansion of PMSM application fields, the requirements for its operational performance are constantly increasing, among which the motor vibration level has become one of the key indicators for evaluating its performance [1,2]. The space vector pulse width modulation (SVPWM) strategy is widely adopted in PMSM control systems due to its excellent dynamic performance [3]. When the SVPWM strategy adopts constant switching frequency modulation, the pulse positions remain relatively fixed, and the conduction timing of the power switching devices in the inverter remains basically unchanged. This results in the high-frequency harmonic components generated during inverter operation being mainly concentrated at the switching frequency and its integer multiples. These high-frequency harmonics are the primary sources of vibration in PMSMs [4,5], which can seriously weaken the operational stability and reliability of PMSM control systems.

Therefore, research on vibration suppression in PMSM control systems has significant theoretical value and practical engineering importance.

To address the aforementioned issues, scholars have conducted extensive research from the perspective of optimizing modulation strategies. The primary focus has been on random pulse width modulation (RPWM) strategies [6,7], periodic pulse width modulation (PPWM) strategies [8,9], and chaos pulse width modulation (CPWM) strategies [10,11]. Among these three spread-spectrum modulation approaches, the PPWM strategy employs periodic functions to vary the switching frequency, which can limit the range of high-frequency harmonic sidebands. However, the spectrum within these sidebands remains discrete, resulting in weak suppression of harmonic peaks. In addition, the types of periodic functions are too single, some periodic functions are difficult to implement, and the spreading effect depends on waveform characteristics, all of which limits the development of the PPWM strategy. The CPWM strategy superimposes chaotic perturbation values onto a traditional fixed switching frequency, causing the inverter's switching frequency to vary chaotically within a certain range. The high-frequency harmonic suppression effect is highly dependent on the selected chaotic map and initial conditions. If the selection is improper, it may lead to unsatisfactory results. Furthermore, the broadband white noise characteristics of chaotic signals in the frequency domain cause the sideband harmonics around the switching frequency to spread across the entire frequency range, generating a large number of low-frequency harmonics. Compared to PPWM and CPWM, RPWM has the characteristics of simple implementation and high modulation signal flexibility. It exhibits good continuity in the frequency domain, effectively avoiding spectral discretization and presenting smoother spectral characteristics. With an excellent spread spectrum performance, RPWM has shown significant advantages in suppressing high-frequency harmonics and optimizing system vibration, and has therefore been widely applied in many fields.

The commonly used RPWM strategies include random switching frequency PWM (RSFPWM) [12,13], random pulse position PWM (RPPPWM) [14,15], random zero-vector PWM (RZVPWM) [16], and dual-random PWM (DRPWM) [17–20]. Ref. [21] proposes an SVPWM strategy based on the Gaussian distribution for random pulse positioning, which reduces the amplitude of high-frequency harmonics by randomly altering pulse positions. However, random numbers generated from a Gaussian distribution exhibit poor randomness, often appearing consecutively on the same side of the expected value. Moreover, a standalone random pulse position strategy still contains impulse functions in its power spectral density, resulting in harmonic peaks in the power spectrum. Ref. [22] introduces a strategy that combines a current harmonic spectrum shaping algorithm with a random switching frequency. Compared with the traditional RSFPWM strategy, it can further reduce the high-frequency harmonic amplitude at the switching frequency and its integer multiples. But it does not fully utilize the degree of freedom of the pulse position in the switching function. Ref. [23] proposes a novel variable-sequence PWM strategy to suppress high-frequency harmonics and current ripple. But this approach requires real-time prediction of current ripple trajectories and the root mean square (RMS) value of current harmonics. Ref. [24] presents a novel SVPWM strategy that eliminates harmonics and noise at odd multiples of the switching frequency by redesigning the switching sequence. However, this reconstruction of switching states increases system switching losses. Ref. [25] proposes a new random zero-vector strategy that reduces the amplitude of high-frequency harmonics by dynamically adjusting the allocation time of zero-vectors. Nevertheless, this standalone random zero-vector strategy has limitations, such as the modulation index increasing, and the effective duration of zero-vectors decreasing. Considering the influence of pulse width and dead time, the spread-spectrum effect gradually weakens. Reference [26] proposes a five-phase dual-random SVPWM strategy for harmonic dispersion at switching frequencies and their harmonics, while generating random numbers by introducing a Beta distribution. However, when optimizing the shape parameters of the Beta distribution, the use of the enumeration method can easily lead to deviations in the optimal shape parameters. Therefore, further optimization can be carried out in the generation of random numbers. The particle swarm optimization (PSO) algorithm, inspired by the foraging behavior of bird flocks [27], is a global optimization algorithm widely used in nonlinear optimization problems due to its simplicity and efficiency. Ref. [2] proposes applying the PSO algorithm to optimize the transition probabilities of Markov chains and random gains, two random parameters in RPWM strategies, to reduce the amplitude of sideband harmonics in motor phase currents, this provides a reason for using the PSO algorithm to optimize random parameters in RPWM strategies. Ref. [28] combines the PSO algorithm for optimizing Beta distribution shape parameters with the long short-term memory (LSTM) model for wind power prediction interval calculation, thereby improving prediction accuracy. However, the method of using the PSO algorithm to optimize Beta distribution shape parameters has not yet been applied to optimizing random PWM strategies.

In summary, to effectively suppress vibration noise and enhance electromagnetic compatibility in PMSM drive systems, this paper proposes an optimized Beta distribution-based dual-random SVPWM strategy. Addressing the issue in conventional SVPWM strategies where a constant switching frequency and zero-vector allocation time lead to concentrated high-frequency harmonics at the switching frequency and its integer multiples, the dual-random SVPWM strategy randomizes the switching frequency and zero-vector allocation time. To overcome the limitations of the conventional dual-random SVPWM strategy, which uses the linear congruential generator (LCG) algorithm and suffers from poor randomness, short periodicity, and fixed distribution, this paper proposes using the adjustable shape and high randomness of the Beta distribution to generate random numbers. Additionally, the PSO algorithm is employed to rapidly optimize the shape parameters of the Beta distribution, avoiding the inefficiency and inaccuracy associated with optimal shape parameter selection in conventional enumeration methods. To verify the effectiveness and feasibility of the proposed strategy, comparative experiments were conducted with the conventional SVPWM strategy and the conventional dual-random SVPWM strategy. The results demonstrate that the proposed improved strategy significantly disperses high-frequency harmonic components in the inverter output signal, effectively reducing high-frequency vibrations in the PMSM.

2. Topology of Two-Level Inverter and Principles of Its Modulation Strategy

The topology of a two-level inverter is shown in Figure 1. In the figure, U_{dc} represents the DC-side voltage; C is the DC-side supporting capacitors; i_A, i_B, and i_C denote the three-phase load currents; and each phase leg consists of two switches, S_{x1} and S_{x2}, where x = A, B, C. Each phase leg has two valid switching states. Taking phase A as an example, the switching states are presented in Table 1.

Table 1. Switching state of phase A.

Switching State	Output State
S_{A1} ON, S_{A2} OFF	1
S_{A1} OFF, S_{A2} ON	0

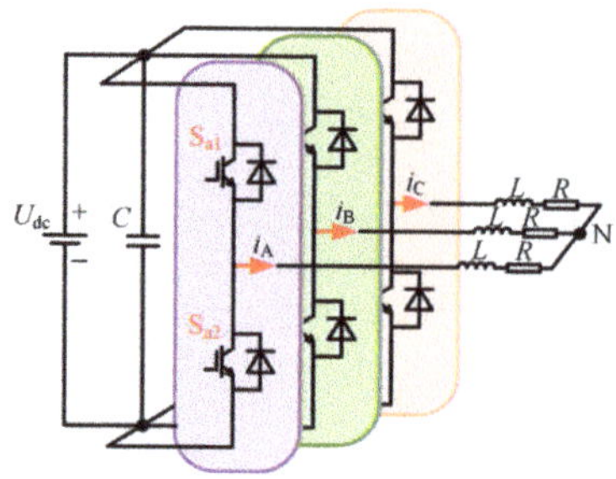

Figure 1. Two-level inverter topology.

From Figure 1 and Table 1, it can be deduced that the inverter has a total of $2^3 = 8$ switching states, corresponding to seven fundamental voltage vectors in the space vector plane, as shown in Figure 2. Among them, the zero-vector V_0 has an amplitude of 0 and corresponds to two switching states; the non-zero vectors V_1 to V_6 have an amplitude of $2U_{dc}/3$ and each corresponds to one switching state.

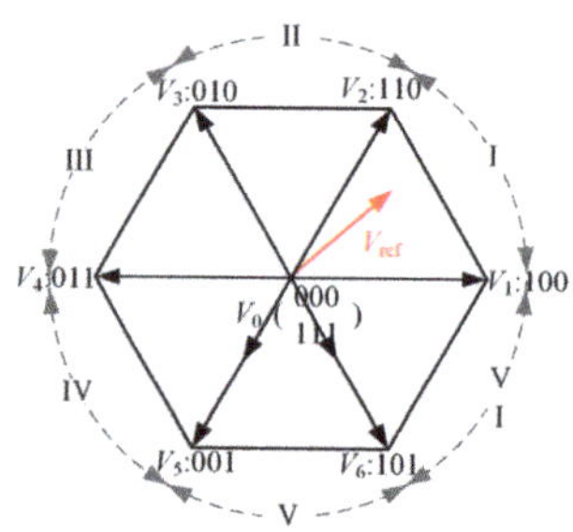

Figure 2. Space vector diagram of a two-level inverter.

The conventional SVPWM strategy is based on the principle of volt second balance, which involves decomposing the product of the reference voltage and the sampling period into the product of the three nearest space vectors and their respective durations. By controlling the switching states of the power devices during the corresponding durations, the three space vectors are output sequentially to achieve modulation. To implement the conventional SVPWM strategy, it is necessary to determine the position of the reference vector and calculate the durations of the fundamental vectors. The durations of each space vector in Figure 2 must satisfy the following conditions:

$$\begin{cases} V_{\text{ref}} T_{\text{s}} = V_x T_x + V_y T_y + V_z T_z \\ T_{\text{s}} = T_x + T_y + T_z \end{cases} \quad (1)$$

where V_{ref} represents the reference voltage vector, T_{s} denotes the sampling period, and V_x, V_y, V_z and T_x, T_y, T_z are the three fundamental vectors closest to the reference voltage vector and their respective durations.

The switching frequency f_{s} of the conventional SVPWM strategy is a fixed value, and the pulses generated by comparing the modulated wave with the carrier are symmetrical about the carrier vertex, which leads to the concentration of high-frequency harmonics in the output signal at the switching frequency and its multiples. The output signal spectrum of the conventional SVPWM strategy is shown in Figure 3.

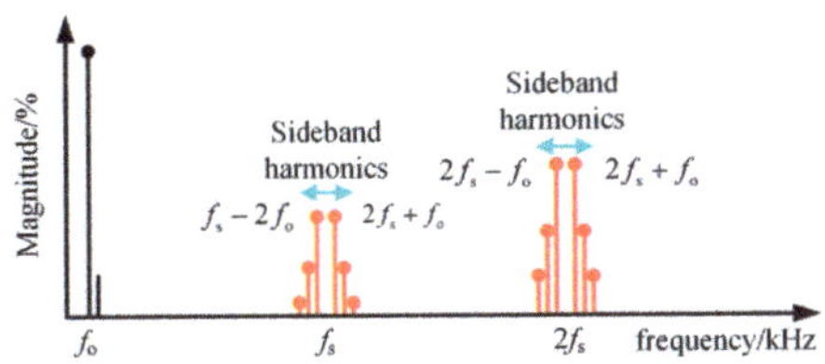

Figure 3. Spectrum diagram of the conventional SVPWM strategy.

3. Conventional Dual-Random SVPWM Strategy

3.1. RZVPWM Strategy

In the conventional SVPWM strategy, the zero-vector duration is uniformly distributed and periodic, which leads to the high-frequency harmonics of the output signal being concentrated at the switching frequency and its integer multiples. However, the duty cycle of the SVPWM strategy is determined solely by the duration of the effective voltage vectors and is independent of the specific allocation method of the zero-vector duration. Therefore, randomizing the allocation of the zero-vector duration can achieve randomization of the pulse conduction positions without affecting the fundamental component of the output voltage. This disrupts the periodic distribution of harmonics in the output signal of the conventional SVPWM strategy, spreading the high-frequency harmonics over a wider frequency range, thereby significantly reducing the harmonic amplitude and improving the high-frequency vibration of the motor. The basic principle of the RZVPWM strategy is illustrated in Figure 4.

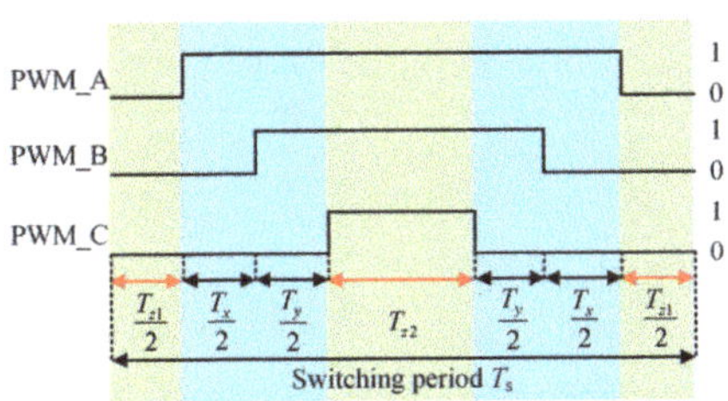

Figure 4. Schematic diagram of the RZVPWM strategy principle.

From Figure 4, the duration of the zero-vector can be expressed as follows:

$$\begin{cases} T_{z1} = R_z T_z \\ T_{z2} = (1 - R_z) T_z \end{cases} \tag{2}$$

where T_z represents the total duration of the zero-vector, T_{z1} denotes the zero-vector duration in the switching state (000), T_{z2} denotes the zero-vector duration in the switching state (111), and R_z is a random number varying between 0.15 and 0.85. By randomizing T_{z1} and T_{z2}, the amplitude of high-frequency harmonics can be reduced.

3.2. RSFPWM Strategy

The RSFPWM strategy builds on the conventional SVPWM strategy by randomizing the switching frequency, thus dispersing the harmonics at the switching frequency and its multiples across a wider frequency domain to achieve the suppression of high-frequency electromagnetic vibrations. The expression for the switching frequency after randomization is given by

$$f_s' = f_s \pm \Delta f \cdot R_f \tag{3}$$

where f_s' represents the randomized switching frequency of the inverter. In this paper, f_s is set to 5 kHz, Δf is set to 1.5 kHz, and R_f is a random number varying between [0, 1]. After specifying the initial value of Δf, the system's switching frequency is altered by randomly changing the value of R_f. Typically, if the inverter's switching frequency is too low, it can easily lead to significant electromagnetic vibrations; conversely, if the switching frequency is too high, it increases the energy loss in the power devices, making the system operation unstable. Therefore, appropriately selecting the variation range of the switching frequency is crucial for ensuring stable system operation. The principle of the RSFPWM strategy is illustrated in Figure 5.

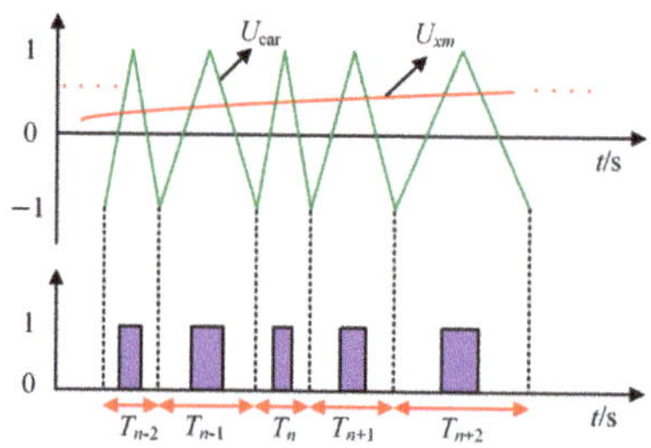

Figure 5. Schematic diagram of the RSFPWM strategy principle.

To address the issue of high-frequency vibration in a PMSM driven by an inverter using the conventional SVPWM strategy, this paper proposes a dual-random SVPWM strategy that combines random zero-vector SVPWM with random switching frequency SVPWM. When applied to the spectral spreading of high-frequency current harmonics, the dual-random SVPWM strategy effectively reduces the motor's high-frequency vibration. The control diagram with the dual-random SVPWM strategy is shown in Figure 6.

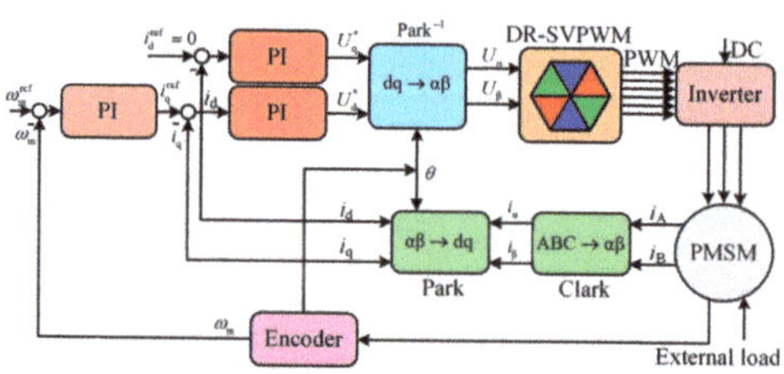

Figure 6. Control diagram with dual-random SVPWM strategy.

When the dual-random SVPWM strategy is applied to motor vibration reduction, the performance of random numbers significantly affects the spread-spectrum effectiveness of the dual-random strategy. The conventional dual-random SVPWM strategy utilizes the LCG algorithm to generate random numbers. The expression for the LCG algorithm is as follows:

$$r_{n+1} = (M_1 \times r_n + M_2)\text{mod}2^{N_s} \tag{4}$$

where r_{n+1} and r_n represent the (n + 1)th and nth generated random numbers, respectively; M_1 and M_2 are two prime numbers; mod denotes the modulo operation; and N_s is the number of bits of the generated random number.

However, when the LCG algorithm is used to generate random numbers, its randomness is limited by the computer's bit precision and the influence of M_1 and M_2, resulting in poor randomness and a short periodicity. These factors constrain the spread-spectrum and vibration reduction effectiveness of the dual-random strategy. Therefore, it is necessary to further optimize the distribution characteristics of the random numbers.

4. Optimization of Random Numbers

4.1. Method Overview

The conventional dual-random SVPWM strategy employs the LCG algorithm to generate random numbers, which exhibit fixed distribution characteristics and lack flexibility, making it impossible to adapt to the specific requirements of harmonic dispersion. This section proposes leveraging the adjustable shape properties of the Beta distribution to generate random numbers in the dual-random SVPWM strategy. Given that the random numbers generated by the Beta distribution are directly influenced by its shape parameters, and to achieve optimal suppression of high-frequency harmonics when these random numbers are applied in the dual-random SVPWM strategy, we propose using the PSO algorithm to rapidly optimize the shape parameters of the Beta distribution, thereby obtaining the best high-frequency harmonic suppression effect.

4.2. The Influence of Different Shape Parameters on the Distribution of Random Numbers

The Beta distribution is a flexible and widely applied continuous probability distribution that can effectively describe the distribution characteristics of variables within the [0, 1] interval. The shape of the Beta distribution is adjustable, making it suitable for generating random numbers that conform to specific distribution shapes. The probability density function of the Beta distribution can be expressed as follows:

$$f(x;a,b) = \frac{x^{a-1}(1-x)^{b-1}}{B(a,b)}, \ 0 \leq x \leq 1 \tag{5}$$

where a and b are shape parameters, both greater than 0, and $B(a, b)$ is the Beta function. By appropriately selecting the shape parameters a and b, Beta distributions with different shapes can be generated. Depending on the values of the shape parameters a and b, the Beta distribution can primarily be categorized into two distinct cases:

(1) When $a = b$, the random number distribution exhibits a high degree of symmetry centered around the expected value of 0.5, a characteristic derived from the mathematical properties of the Beta distribution under parameter symmetry.

(2) When $a \neq b$, the random number distribution displays asymmetry, with its probability density function becoming imbalanced near the expected value of 0.5, skewing toward the side with the larger parameter.

In random PWM strategies, the distribution of random numbers should ideally be symmetrical about the expected value of 0.5. When the distribution of random numbers is asymmetric, it can cause a significant deviation in the overall average switching frequency of the system, manifesting as either an excessively high or low average switching frequency. A higher average switching frequency increases the switching losses of power devices, thereby reducing the system's energy efficiency. Conversely, a lower average switching frequency exacerbates harmonic distortion in the output voltage and current, degrading the waveform quality of the output signal and overall system performance. Both scenarios contradict the original intent of the random strategy, which aims to reduce the amplitude of high-frequency harmonics through randomized switching frequencies. Therefore, this study focuses on the case where the shape parameter $a = b$. Leveraging its symmetric distribution characteristics, the application effects and optimization potential of this case in the dual-random SVPWM strategy are thoroughly investigated to ensure that the system's efficiency and performance meet the expected objectives.

Figure 7 illustrates the Beta distribution under different shape parameters. As shown in Figure 7, when $a = b = 1.5$, the Beta distribution curve exhibits a convex shape symmetrical about the expected value of 0.5. In this case, as the random number approaches the expected value of 0.5, the probability density of the random number increases, and vice versa. For $a = b > 1$, as the values of a and b increase, the probability density of random numbers near the expected value of 0.5 also increases. When $a = b = 1$, the Beta distribution curve becomes a straight line with a constant probability density of 1, indicating that all random numbers have an equal probability density, and the random numbers follow a uniform distribution. When $a = b < 1$, the Beta distribution curve takes on a concave shape symmetrical about the expected value of 0.5. Here, as the random number approaches the expected value of 0.5, the probability density decreases, and vice versa. For $a = b < 1$, as the values of a and b decrease, the probability density of random numbers near the expected value decreases. It is evident that different shape parameters directly influence the distribution characteristics of random numbers. Therefore, to achieve the optimal spread-spectrum vibration reduction effect when applying the dual-random SVPWM strategy, optimizing the selection of shape parameters is crucial.

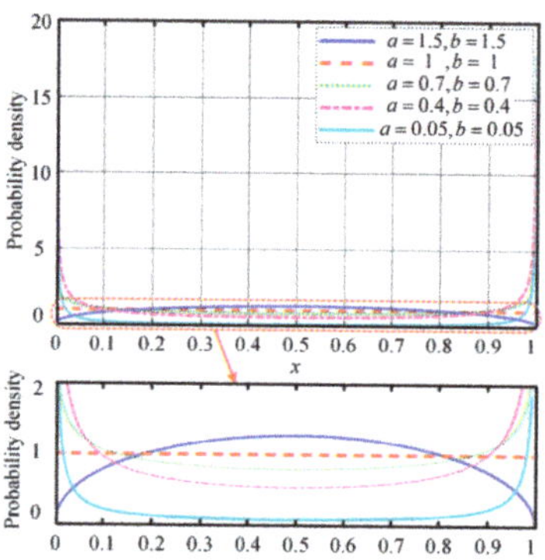

Figure 7. Beta distribution curves under different shape parameters.

4.3. *Shape Parameter Optimization Based on PSO Algorithm*

To evaluate the effectiveness of the dual-random SVPWM strategy proposed in this paper, the power spectral density (PSD) plot and the HSF are utilized to analyze the impact of different shape parameters on the spread-spectrum effect.

The PSD function of the dual-random SVPWM strategy can be expressed as

$$S_D(\omega, R_T, R_\varepsilon) = \frac{1}{E[T_s]}\left\{E\left[|F(\omega, R_T, R_\varepsilon)|^2\right] + 2\mathrm{Re}\left[\frac{E[F^*(\omega, R_T, R_\varepsilon)] \cdot E\left[F^*(\omega, R_T, R_\varepsilon)e^{j\omega T_s}\right]}{1 - E\left[e^{j\omega T_s}\right]}\right]\right. \quad (6)$$

where R_T is the reciprocal of R_f, R_ε is the pulse delay coefficient corresponding to the random zero vector, F is the Fourier transform of the switching function, F^* represents the complex conjugate of F, T_s is the switching period, and E represents the expectation operator.

The PSD plot, as a critical visualization tool for frequency-domain analysis, can accurately characterize the distribution of signal energy across the frequency axis, thereby providing an intuitive representation of the spectral broadening effect and smoothness achieved by the random strategy during the spread-spectrum process. In contrast to the qualitative analysis provided by the PSD plot, the HSF quantitatively assesses the spread-spectrum effect of the random strategy by calculating the standard deviation of the harmonic amplitudes in the sample data. A smaller HSF indicates a better harmonic dispersion effect of the employed modulation strategy. The HSF can be expressed as

$$HSF = \sqrt{\frac{1}{N}\sum_{j>1}^{N}(H_j - H_0)^2} \tag{7}$$

$$H_0 = \frac{1}{N}\sum_{j>1}^{N} H_j \tag{8}$$

$$H_j = \sqrt{\int_0^1 \int_0^1 f(R_z; a, b) f(R_f; a, b) \left| I(j \cdot f_0, R_z, R_f) \right|^2 dR_z dR_f} \tag{9}$$

where H_0 is the average amplitude of all (N) harmonics excluding the fundamental wave, H_j represents the amplitude of the j-th harmonic, and I ($j{\cdot}f_0$, R_z, R_f) represents the Fourier transform of the phase current i (t) at the frequency, with f_0 denoting the spectral resolution. The factors influencing H_j include R_z, R_f, a, b, j, f_0, and the motor speed. However, when comparing the j-th harmonic amplitudes across different strategies, j, f_0, and the motor speed should be controlled as identical variables. When random numbers generated by the LCG algorithm and the Beta distribution are applied to R_z and R_f, they will affect H_j.

Since different shape parameters correspond to different Beta distributions, the performance of random numbers generated using the Beta distribution will be affected when applied. To compare the spread-spectrum effects of random numbers under different shape parameters, the PSD plots from simulations are first used for comparison. Taking the shape parameters $a = b = 0.5$, $a = b = 1$, and $a = b = 1.5$ with a speed of 1200 rpm as an example, Figure 8 illustrates the PSD plots of the simulated phase currents for the dual-random SVPWM strategy under these three shape parameter sets.

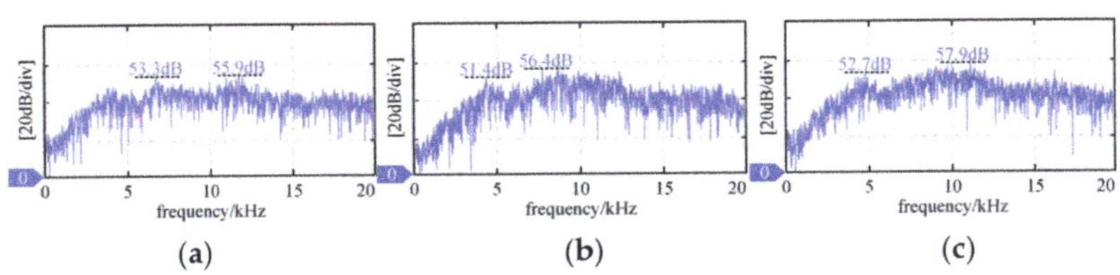

Figure 8. PSD diagram under different shape parameters: (**a**) $a = b = 0.5$. (**b**) $a = b = 1$. (**c**) $a = b = 1.5$.

From the figure above, it can be observed that the harmonic amplitudes at 5 kHz and 10 kHz under the three cases of the dual-random SVPWM strategy exhibit significant differences. This indicates that the performance of random numbers following a Beta distribution largely depends on the choice of shape parameters. The selection of different shape parameters ultimately affects the spread-spectrum effect.

Since the HSF can quantitatively analyze the harmonic spread-spectrum effects of different shape parameters, it is utilized as the evaluation objective function in this study. On this basis, the PSO algorithm is employed to rapidly optimize the shape parameters, aiming to achieve the minimum HSF and thereby obtain the optimal suppression effect on high-frequency vibrations.

The PSO algorithm is a swarm intelligence optimization algorithm inspired by the collective foraging behavior of bird flocks, classified as a type of metaheuristic optimization method. Unlike traditional algorithms such as Simulated Annealing (SA) and the Genetic Algorithm (GA), PSO does not rely on population crossover or mutation operations. Instead, it iteratively approaches the optimal solution through internal interactions among particles. Owing to its characteristics of global high-precision convergence and robust reliability, PSO has been widely applied in fields such as multi-objective optimization, adaptive control, nonlinear problems, and multidimensional space optimization.

As a random search algorithm, the core driving mechanism of the PSO algorithm involves the shared iterative updating of the global historical best solution G_{best} and the individual historical best solution P_{best}. This process enables the optimization of both individual extrema and the global optimum of the particle swarm. In this paper, each particle represents a candidate solution vector P (a_i, b_i) for the shape parameters of the Beta distribution, where $a_i = b_i$. The particle's position defines the distribution characteristics of random numbers R_z and R_f. The objective function is the HSF, and optimizing a_i and b_i minimizes the HSF, thereby reducing the high-frequency harmonic peak. The specific workflow of the PSO algorithm is illustrated in Figure 9.

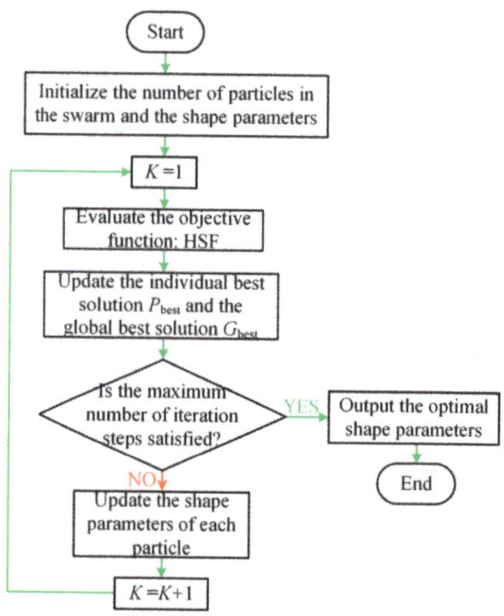

Figure 9. Flowchart of the PSO algorithm.

The screening and solving process for the global optimal solution G_{best} is as follows:

(1) Calibration of initial system parameters: Based on the phase current data collected from experiments, calibrate parameters such as winding inductance and resistance in the simulation model to ensure that the simulation model aligns with experimental results under various operating conditions.

(2) Initialize the particle swarm: Restrict the particle range to 0.01~2. Each particle in the swarm contains basic information, namely the shape parameters a and b. During each iteration, the individual best solution P_{best} of each particle is compared and updated with the global best solution G_{best}. Since this study involves an optimization problem with only two parameters, a population of 20 particles is sufficient to cover the solution space, providing adequate diversity to avoid premature convergence to suboptimal solutions. Additionally, the maximum number of iterations is chosen as the convergence criterion, with 60 iterations allowing the algorithm sufficient time to refine the solution. A dynamic inertia weight strategy is adopted, with the weight value linearly decreasing from 0.9 to 0.4 during the iteration process. A larger weight in the early stages ensures a strong global search capability, while a smaller weight in the later stages enhances the local search capability. The learning factors are set to $c1 = c2 = 2.0$, achieving a balance between individual experience and collective collaboration, thus preventing premature convergence to local optima.

(3) Run the steady-state condition Simulink simulation program, generate the phase current time-domain waveform data in the MATLAB (Version: 9.5.0.944444, R2018b) (MathWorks, Inc., Natick, MA, USA) workspace, and then perform time-frequency conversion on the data using the fast Fourier transform (FFT) program, followed by calculating the HSF, and finally, extract the individual optimal solution P_{best} and the global optimal solution G_{best}.
(4) Termination condition setting: In the conventional PSO algorithm, termination conditions typically include the number of iteration steps and convergence criteria. To ensure population diversity, this paper uses the number of iteration steps as the termination condition to prevent premature convergence or excessive iteration without convergence of the particle swarm.

The optimization results of the Beta distribution shape parameters based on the PSO algorithm are shown in Figure 10. The HSF stabilizes within 46 iterations, and as the number of iterations continues to increase, the HSF no longer decreases. At this point, the shape parameters a and b are both equal to 0.68, with an HSF of 1.796. Therefore, under these shape parameters, the random numbers generated by the Beta distribution, when applied to the dual-random SVPWM strategy, achieve the optimal spectrum-spreading effect.

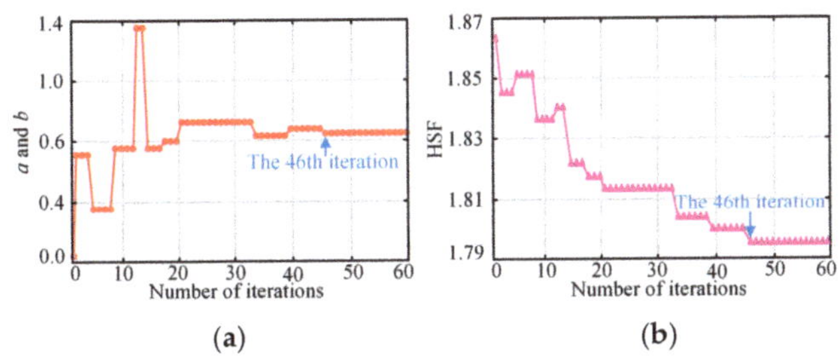

Figure 10. Optimization results of the PSO algorithm: (**a**) a and b after iteration. (**b**) HSF after iteration.

Furthermore, Figure 11a,b, respectively, depict an autocorrelation function (ACF) graph and absolute autocorrelation function (|ACF|) graph comparing the randomness of random numbers generated by the Beta distribution and the LCG algorithm.

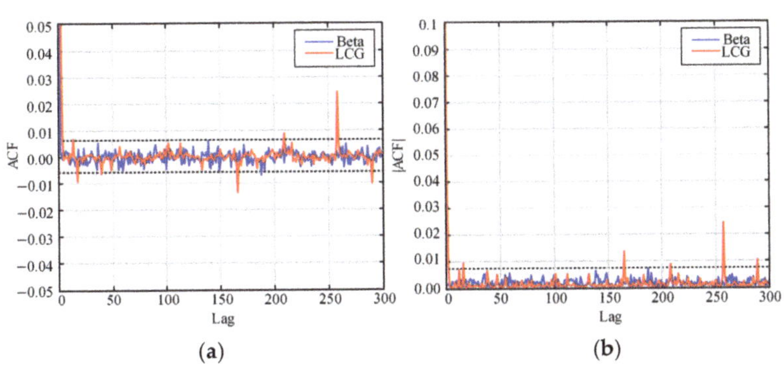

Figure 11. ACF graph and |ACF| graph of different random number algorithms: (**a**) ACF graph. (**b**) |ACF| graph.

From the ACF graph, it can be observed that the random numbers generated by the Beta distribution exhibit an ACF with very small fluctuations across all lag points, with values close to 0 and consistently within the confidence interval (indicated by dashed lines), suggesting that the Beta distribution has almost no autocorrelation; in contrast, the random numbers generated by the LCG algorithm also show an ACF close to 0 at most lags, but they exhibit multiple significant peaks (e.g., around lags 50, 150, and 250), which exceed the confidence interval, indicating clear periodic characteristics.

From the | ACF | graph, it can be observed that the random numbers generated by the Beta distribution exhibit consistently low | ACF | values, fluctuating between 0 and 0.01, indicating extremely low autocorrelation strength. In contrast, the random numbers generated by the LCG algorithm also show low | ACF | values at most lags, but there are several significant peaks (e.g., around lags 50, 150, and 250), corresponding to the ACF peaks observed in the left graph.

Table 2 compares the mean absolute autocorrelation coefficient (MAACF), maximum absolute autocorrelation coefficient (MaxAACF), and the number of significant non-zero lags (NSNLs) of the random numbers generated by these two algorithms.

Table 2. Comparison of randomness metrics for Beta distribution and LCG algorithm.

Algorithm	MAACF	MaxAACF	NSNL
Beta distribution (a = b = 0.68)	0.001767	0.007074	2
LCG algorithm	0.002116	0.199811	10

From the table above, it can be observed that the random numbers generated by the Beta distribution have lower values in MAACF = 0.001767, MaxAACF = 0.007074, and NSNL = 2 compared to those generated by the LCG algorithm (MAACF = 0.002116, MaxAACF = 0.199811, NSNL = 10), further indicating that the random numbers generated by the Beta distribution exhibit superior randomness compared to the LCG algorithm.

5. Experimental Results

This paper takes a 4.3 kW PMSM as the research object. To verify the feasibility and superiority of the proposed dual-random SVPWM strategy based on an optimized Beta distribution, experimental validation and analysis were conducted. The proposed strategy was compared with the conventional SVPWM strategy and the conventional dual-random SVPWM strategy in terms of harmonic amplitude and vibration acceleration peak values at different frequencies. The experimental platform is illustrated in Figure 12. This experiment utilized the RT-LAB platform (OPAL-RT Technologies Inc., Montreal, QC, Canada) to implement the system control algorithm, with a two-level inverter serving as the power circuit. Current and voltage data were collected using a Yokogawa high-performance signal analysis oscilloscope, while vibration acceleration signals were acquired using an accelerometer and a Siemens multifunctional data acquisition device. Table 3 classifies different strategies.

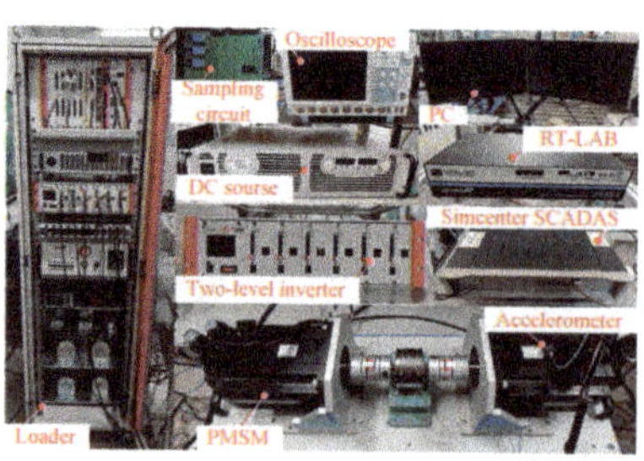

Figure 12. Experimental platform.

Table 3. Classification of different random strategies.

Strategy	Principle
Strategy I	Conventional SVPWM strategy
Strategy II	Dual-random SVPWM strategy based on the LCG algorithm
Strategy III	Dual-random SVPWM strategy based on an optimized Beta distribution

The experimental system parameters are presented in Table 4. The experiments in this study used 300 rpm and 1200 rpm as test speeds based on the following considerations: 300 rpm represents low-speed operation, commonly observed in low-speed cruising for electric vehicles or low-speed servo scenarios in industrial automation; 1200 rpm is close to the rated speed (Table 4 in the paper indicates a rated speed of 1500 rpm), representing medium-to-high-speed operation. Furthermore, as evidenced by the PSD plots from the simulations, Strategy III demonstrates superior harmonic suppression even under other speed conditions. By combining these aspects, the selection of these two speeds effectively reflects the low-speed and high-dynamic-response performance of PMSM applications.

Table 4. Experimental system parameters.

Parameter	Value
Rated voltage/V	220
Rated current/A	20
Rated speed/rpm	1500
V_{DC} bus voltage/V	350
Fixed switching frequency/kHz	5
Random switching frequency variation range/kHz	3.5~6.5

Figures 13 and 14 show the PSD plots of the phase currents for the three strategies at different speeds. To facilitate a comparison between the strategies, the harmonic amplitudes at 5 kHz and 10 kHz frequencies are summarized in Table 5. At a speed of 300 rpm, at the 5 kHz frequency, Strategy II reduces the harmonic amplitude by 9.5 dB compared to Strategy I, while Strategy III further reduces it by 1.9 dB compared to Strategy II; at the 10 kHz frequency, Strategy II reduces the harmonic amplitude by 23.9 dB compared to Strategy I, and Strategy III further reduces it by 3.7 dB compared to Strategy II. At a speed of 1200 rpm, comparing the harmonic amplitudes at 5 kHz, Strategy II achieves a reduction of 15.6 dB compared to Strategy I, with Strategy III further reducing it by 4.5 dB compared to Strategy II. At 10 kHz, Strategy II reduces the harmonic amplitude by 22 dB compared to Strategy I, and Strategy III further reduces it by 1.8 dB compared to Strategy II. Thus, it can be concluded that, across different speeds, Strategy III exhibits the lowest harmonic amplitudes at both 5 kHz and 10 kHz.

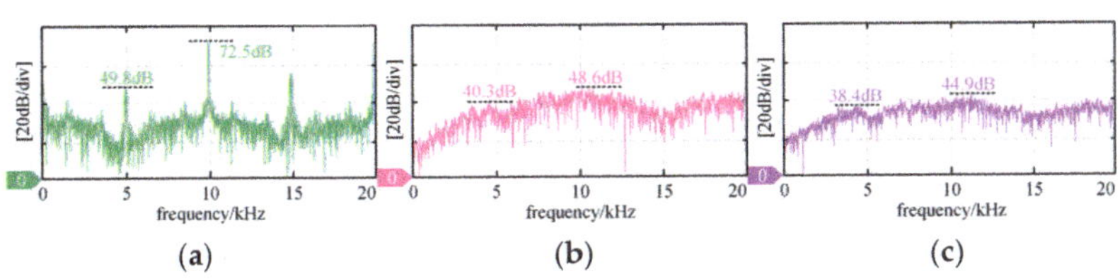

Figure 13. PSD graphs of phase currents for three strategies at 300 rpm: (**a**) Strategy I. (**b**) Strategy II. (**c**) Strategy III.

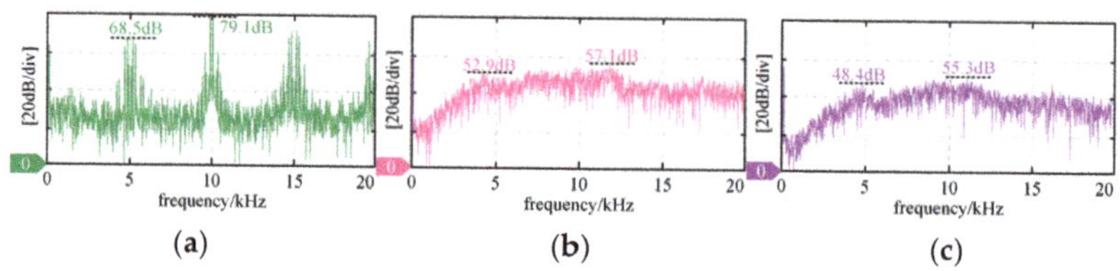

Figure 14. PSD graphs of phase currents for three strategies at 1200 rpm: (**a**) Strategy I. (**b**) Strategy II. (**c**) Strategy III.

Table 5. Comparison of harmonic amplitudes at different speeds.

Strategy	300 rpm		1200 rpm	
	5 kHz	10 kHz	5 kHz	10 kHz
Strategy I	49.8 dB	72.5 dB	68.5 dB	79.1 dB
Strategy II	40.3 dB	48.6 dB	52.9 dB	57.1 dB
Strategy III	38.4 dB	44.9 dB	48.4 dB	55.3 dB

Figures 15 and 16 illustrate the vibration acceleration plots for the three strategies at different speeds. To facilitate comparison between the strategies, the peak vibration acceleration values at 5 kHz and 10 kHz frequencies are summarized in Table 6. At a speed of 300 rpm, at the 5 kHz frequency, Strategy II reduces the peak vibration acceleration by 0.0174 g compared to Strategy I, while Strategy III further reduces it by 0.0004 g compared to Strategy II; at the 10 kHz frequency, Strategy II reduces the peak vibration acceleration by 0.0691 g compared to Strategy I, and Strategy III further reduces it by 0.0063 g compared to Strategy II. At a speed of 1200 rpm, at the 5 kHz frequency, Strategy II reduces the peak vibration acceleration by 0.0386 g compared to Strategy I, with Strategy III further reducing it by 0.0008 g compared to Strategy II; at the 10 kHz frequency, Strategy II reduces the peak vibration acceleration by 0.1732 g compared to Strategy I, and Strategy III further reduces it by 0.0085 g compared to Strategy II. Therefore, it can be concluded that, across different speeds, Strategy III consistently exhibits the lowest peak vibration acceleration at both 5 kHz and 10 kHz.

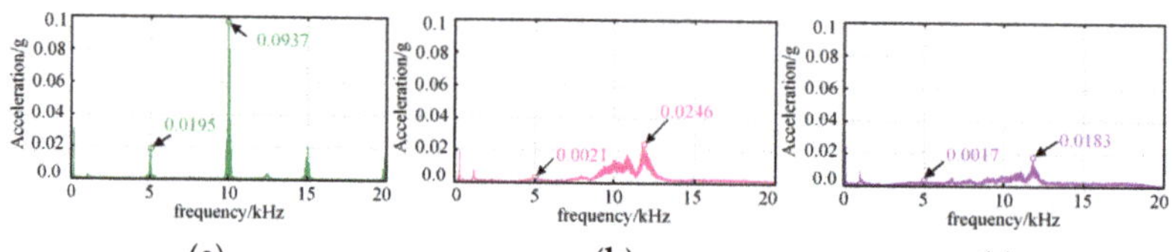

Figure 15. High-frequency vibration acceleration graphs for three strategies at 300 rpm: (**a**) Strategy I. (**b**) Strategy II. (**c**) Strategy III.

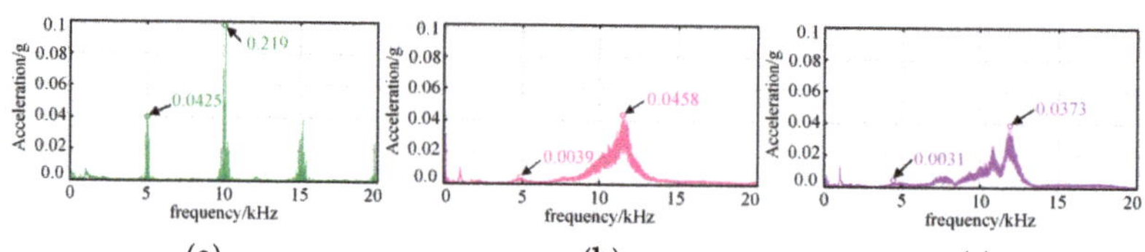

Figure 16. High-frequency vibration acceleration graphs for three strategies at 1200 rpm: (**a**) Strategy I. (**b**) Strategy II. (**c**) Strategy III.

Table 6. Comparison of vibration acceleration peaks at different speeds.

Strategy	300 rpm		1200 rpm	
	5 kHz	10 kHz	5 kHz	10 kHz
Strategy I	0.0195 g	0.0937 g	0.0425 g	0.2190 g
Strategy II	0.0021 g	0.0246 g	0.0039 g	0.0458 g
Strategy III	0.0017 g	0.0183 g	0.0031 g	0.0373 g

As Strategy I employs a constant switching frequency and fixed zero-vector allocation modulation, pulse positions are relatively fixed. This causes the high-frequency harmonic components generated during inverter operation to form clustered harmonic spikes at the switching frequency and its multiples. Strategy II uses the LCG algorithm to generate random numbers. However, the random numbers generated by the LCG algorithm exhibit poor randomness and short periodicity, and they cannot be adjusted based on specific harmonic dispersion requirements. Consequently, although Strategy II introduces a degree of randomness, its harmonic dispersion effect remains limited, achieving only a moderate reduction in harmonic peaks, while harmonic spikes persist.

Compared to Strategy I, this paper proposes randomizing the switching frequency and zero-vector allocation time to avoid high-frequency harmonics caused by a fixed switching frequency and fixed zero-vector allocation time. Compared to Strategy II, this paper proposes using the Beta distribution to generate random numbers, addressing the issues of poor randomness and short periodicity in random numbers generated by the LCG algorithm. Additionally, it combines the PSO algorithm to rapidly optimize the shape parameters of the Beta distribution, improving the quality and generation efficiency of random numbers. Consequently, compared to Strategies I and II, Strategy III is optimal, achieving the lowest harmonic amplitudes at the switching frequency and its multiples, and minimizing the peak vibration acceleration of the permanent magnet synchronous motor.

Theoretically, increasing Δf enhances high-frequency harmonic dispersion but tends to introduce low-frequency harmonics, leading to distortion in the inverter's output line voltage, increased current fluctuations, and potential EMC compliance issues. Conversely, reducing Δf weakens the dispersion effect but improves control precision. Variations in R_z also affect harmonic dispersion: a smaller R_z range reduces dispersion capability, while excessive randomness in R_z increases waveform fluctuations, impacting precision in industrial automation applications.

From the above analysis, it is evident that, compared to the conventional dual-random SVPWM strategy, the use of the dual-random SVPWM control based on an optimized Beta distribution in a PMSM can effectively achieve the effects of spectrum spreading and vibration reduction.

6. Discussion

In this study, the PSO algorithm is executed offline. Once the optimal parameters are determined, they are fixed and hard-coded into the control system. Consequently, there is no need to run the PSO algorithm in real-time during motor operation. For random number sampling from the Beta distribution, this study employs the inverse transform sampling method, which can be efficiently implemented in real-time systems and is computationally feasible for embedded systems. By pre-computing a lookup table for the cumulative distribution function (CDF), random numbers can be rapidly generated during real-time operations. Furthermore, the FFT and HSF calculations in this study are primarily used for offline analysis and evaluation and are not part of the real-time control loop. During real-time operation, the control algorithm relies solely on pre-optimized shape parameters and random number generation, without requiring FFT or HSF computations.

Although this study employs the PSO algorithm to optimize the shape parameters of the Beta distribution, with the optimized parameters subsequently hard-coded into the system—a process that does not directly impact real-time response time—the offline optimization process itself demands significant computational resources and high-performance computing equipment, thereby increasing development costs. The PSO algorithm involves multiple iterations (60 iterations with 20 particles in this study), each requiring Simulink simulations, FFT computations, and HSF calculations, which may pose challenges for resource-constrained research environments or small development teams.

The offline PSO algorithm for optimizing Beta distribution shape parameters involves 60 iterations with 20 particles, each requiring Simulink simulations, FFT, and HSF calculations, necessitating high-performance computing resources but not affecting real-time operations. Real-time computations encompass Beta distribution random number generation, utilizing the inverse transform sampling method to compile a pre-computed cumulative distribution function lookup table for generating random parameters R_z and R_f, which consumes control loop time and may introduce delays. Additionally, dynamic PWM adjustments for the switching frequency and zero-vector allocation time increase the computational burden of the control loop, potentially interfering with other tasks. These factors impact the response time of the PMSM control system through computational delays and timing conflicts. For field-oriented control (FOC) and direct torque control (DTC), the delays are typically acceptable. However, for model predictive control (MPC), delays may lead to accumulated prediction errors, requiring compensation mechanisms. Furthermore, timing conflicts can cause sampling window offsets, compromising the accuracy of feedback signals.

The concentration of high-frequency harmonics is a primary source of electromagnetic interference (EMI). The strategy proposed in this study randomizes the switching frequency (3.5–6.5 kHz) and zero-vector allocation time (based on random numbers R_z and R_f), dispersing the high-frequency harmonics, which are typically concentrated at the switching frequency and its multiples (e.g., 5 kHz, 10 kHz) in conventional SVPWM, across a broader spectrum. This dispersion reduces the energy at individual frequency points, thereby mitigating electromagnetic radiation and conducted interference, and enhancing the system's electromagnetic compatibility (EMC). Compared to the random numbers generated by the conventional LCG algorithm, which suffer from poor randomness and short periodicity, the high-quality random numbers produced by the Beta distribution (with optimized parameters a = b = 0.68) through the PSO algorithm ensure more uniform spectrum spreading. The superior randomness of the Beta distribution is validated in this study through the autocorrelation function (ACF) and absolute autocorrelation coefficients (MAACF = 0.001767, MaxAACF = 0.007074), which reduce periodic harmonic peaks and further lower EMI risks. However, while high-frequency harmonics are effectively dispersed, the randomization process may introduce low-frequency harmonics. In this study, the proposed dual-random SVPWM strategy exhibits high stability under light load conditions, attributed to low current demands and effective high-frequency harmonic dispersion, making it suitable for applications such as small servo motors and fan drives. However, stability may be compromised under high load conditions due to increased current harmonics and response lag induced by computational delays, necessitating self-adaptive parameter correction and overload protection to enhance fault tolerance. By implementing control optimization, fault detection algorithms, and parameter adaptation, issues related to low-frequency harmonics, delays, and fault diagnosis can be mitigated, ensuring the strategy's robustness and reliability across industrial applications ranging from low-power to high-performance scenarios.

The proposed solution in this study primarily achieves high-frequency harmonic dispersion by modifying the PWM signal generation module, which can be integrated as an independent module into existing PMSM control systems. The PSO algorithm optimizes the Beta distribution shape parameters offline and hard-codes them, with real-time operations limited to random number generation and PWM adjustments, decoupled from upper-level control logic (such as FOC, DTC, and MPC), thereby reducing integration complexity. Modern PMSM control systems, typically based on high-performance DSPs, FPGAs, or MCUs, possess the capability to support dynamic PWM adjustments, store CDF lookup tables, and handle additional computational loads. However, PWM modules in low-end systems may lack support for dynamic frequency adjustments or sufficient clock precision, resulting in modulation errors. The CDF lookup table requires 10–100 KB of storage space, which may exceed the memory capacity of low-end MCUs, necessitating external Flash storage or table compression. Hardware incompatibility in low-end MCUs may lead to signal distortion, thereby reducing the effectiveness of harmonic dispersion.

7. Conclusions

To address the high-frequency vibration issue of PMSM, this paper proposes an optimized Beta distribution-based dual-random SVPWM strategy to suppress high-frequency harmonics and reduce motor high-frequency vibrations. The advantages of the proposed method are summarized as follows:

(1) To address the high-frequency vibration issue in conventional SVPWM strategies caused by a constant switching frequency and fixed zero-vector allocation time, randomization of the switching frequency and zero-vector allocation time is proposed to reduce high-frequency vibrations in PMSM.
(2) To address the issues of poor randomness and short periodicity of random numbers generated by the LCG algorithm in conventional dual-random SVPWM strategies, the use of the Beta distribution for random number generation is proposed to improve the quality of random numbers.
(3) To address the inefficiency of using enumeration methods to find optimal shape parameters in traditional Beta distribution-based random number generation, a PSO algorithm is proposed to rapidly optimize shape parameters, thereby improving efficiency.

In summary, the proposed method not only improves the efficiency and quality of random number generation but also further suppresses high-frequency harmonics in phase currents, thereby reducing high-frequency vibrations in PMSM.

Author Contributions: Conceptualization, G.Z. and X.G.; methodology, G.Z. and K.W.; software, K.W.; validation, K.W., G.Z. and X.J.; formal analysis, G.Z. and X.G.; writing—original draft preparation, K.W.; writing—review and editing, X.G., X.J. and C.L.; funding acquisition, W.C. and X.G. All authors have read and agreed to the published version of the manuscript.

Funding: This research was funded by the Pioneer Project of Zhejiang Province, grant number 2024C01014, and the National Natural Science Foundation of China, grant number 52177055.

Data Availability Statement: Data are contained within the article.

Conflicts of Interest: The authors declare no conflicts of interest.

References

1. Jang, I.S.; Kim, W.H. Study on Electromagnetic Vibration Analysis Process for PM Motors. IEEE Transactions on Applied Superconductivity. *IEEE Trans. Appl. Supercond.* **2020**, *30*, 1–6.
2. Chen, Y.; Qiu, Z.Z.; Ma, K.; Kong, Z.G.; Huang, X. Investigation into Markov-chain random modulation for suppression high-frequency sideband vibro-acoustics in permanent magnet synchronous motor. *Electr. Mach. Control.* **2023**, *27*, 109–118.

3. Dong, L.J.; Pan, Z.L.; Li, S.; Li, Y.; Duan, H. Micromotors Study on Regulation and Voltage Utilization Ratio of SPWM and SVPWM Schemes for Vehicle Motor. *Micromotors* **2023**, *56*, 49–52.
4. Li, F.T.; Zhang, C.X.; Pang, Y. Analysis and Optimization of Electromagnetic Vibration and Noise of Permanent Magnet Synchronous Motor. *Small Spec. Electr. Mach.* **2024**, *52*, 27–32.
5. Li, Y.Z.; Wu, S.N. Methods for Vibration and Noise Reduction of IPMSM. *Small Spec. Electr. Mach.* **2024**, *52*, 25–29.
6. Xu, J.; Ouyan, Z.; Sun, J.; Zhu, W.; Nie, Z. Performance and Characterization of Optimal Harmonic Dispersion Effect in Double Frequency-Band Random PWM Strategy. *IEEE Trans. Power Electron.* **2024**, *39*, 14680–14690. [CrossRef]
7. Zhang, P.; Wang, S.; Li, Y. Generalized N-State Random Pulse Position Discontinuous PWM for High-Frequency Harmonics Reduction and Performance Improvement at High Modulation Ratios. *IEEE Trans. Power Electron.* **2024**, *39*, 13659–13671. [CrossRef]
8. Qiu, Z.Z.; Chen, Y.; Cheng, H.Q.; Li, X.; Gu, F.S. Periodic Harmonic Spread Spectrum Modulation for High-Frequency Sideband Vibro-Acoustic Suppression in Permanent Magnet Synchronous Motor. *Trans. China Electro. Soc.* **2022**, *37*, 2459–2468.
9. Ji, Z.; Cheng, S.; Li, X.; Lv, Y.; Wang, D. An Optimal Periodic Carrier Frequency PWM Scheme for Suppressing High-Frequency Vibrations of Permanent Magnet Synchronous Motors. *IEEE Trans. Power Electron.* **2023**, *38*, 13008–13018. [CrossRef]
10. Natarajan, S.; Babu, T.S.; Balasubramanian, K.; Sain, C.; Satpathy, P.R. An Effective EMI Mitigation Technique Using Chaotic PWM for Interleaved Boost Converter. In Proceedings of the 2021 International Conference in Advances in Power, Signal, and Information Technology (APSIT), Bhubaneswar, India, 8–10 October 2021.
11. Li, H.; Li, Y.; Wang, Z.; Zhou, M.; Zhang, B.; Yang, Z. Logistic-Chebyshev Cascaded Chaotic Space Vector Pulse Width Modulation for Common-Mode EMI Suppression in Motor Drive System. In Proceedings of the 2024 2nd China Power Supply Society Electromagnetic Compatibility Conference (CPEMC), Hangzhou, China, 16–18 August 2024.
12. Bu, F.F.; Ma, B.J.; Du, R.H.; Sun, T.K.; Chu, J.B.; Liu, Q. Multiaverage Random Switching Frequency Space Vector Pulsewidth Modulation Strategy for High-Order Harmonics Dispersion. *IEEE J. Emerg. Sel. Topics Power Electron.* **2023**, *11*, 4010–4021. [CrossRef]
13. Wen, J.; Cheng, X.; Gao, Y.; Liu, J.; Ji, P.; Yang, J. An Optimization Method in the Random Switching Frequency SVPWM for the Sideband Harmonic Dispersion. *IEEE J. Emerg. Sel. Topics Power Electron.* **2024**, *12*, 4801–4813. [CrossRef]
14. Zhang, P.; Wang, S.; Li, Y. Three-Phase Two-Level VSIs With Significant PWM Harmonics Dispersion and Improved Performance Using Generalized N-State Random Pulse Position SVPWM With Constant Sampling Frequency. *IEEE Trans. Power Electron.* **2024**, *39*, 1394–1409. [CrossRef]
15. Deng, W.; Huang, J.; Qian, Z.; Qian, C.; Zhong, D. A Random Pulse Position-Based Selective Noise Cancellation Modulation Method for SVPWM Driven PMSMs. *IEEE Trans. Energy Convers.* **2022**, *37*, 2190–2198. [CrossRef]
16. Li, S.S.; Huang, L.F.; Wang, W.S.; Li, X.; Han, X. Effect of Two Zero Voltage Vectors Action Time on Common-mode and Harmonic Components of Output Voltage for Indirect Matrix Converter. *Proc. Chin. Soc. Electr. Eng.* **2022**, *42*, 7943–7955.
17. Wang, Y.; Liu, J.; Lu, B.; Wang, M. A Novel Discrete Hybrid Dual Random SVPWM Scheme for Reducing PMSM Harmonic Intensity. *IEEE/ASME Trans. Mechatron.* **2023**, *28*, 1425–1435. [CrossRef]
18. Zhang, P.; Wang, S.; Li, Y. A Novel Dual Random Scheme in Signal Injection Sensorless Control of IPMSM Drives for High-Frequency Harmonics Reduction. *IEEE Trans. Power Electron.* **2023**, *38*, 14450–14462. [CrossRef]
19. Zhang, P.; Wang, S.; Li, Y. Sensorless Control of IPMSM in Zero- and Low-Speed Regions Using Dual Random High-Frequency Square-Wave Voltage Injection to Reduce Audible Noise. *IEEE Trans. Ind. Electron.* **2024**, *71*, 11864–11875. [CrossRef]
20. Xu, J.; Ouyang, Z.; Dong, Y.; Nie, Z. Spectrum-Dispersion Effects of Random Triangular Carrier Wave SVPWM for NPC Three-Level Inverters. *IEEE Trans. Ind. Electron.* **2024**, *71*, 7756–7765. [CrossRef]
21. Wu, W.M.; Li, G.L.; Xie, F.; Liang, K.K.; Qi, C.M. Harmonic Suppression Strategy of SVPWM with Random Pulse Position Based on Gauss Distribution. *Micromotors* **2020**, *53*, 50–55+89.
22. Liu, H.P.; Liu, Q.; Zhang, W.; Miao, Y.R.; Li, P. Random PWM Technique for Acoustic Noise and Vibration Reduction in Induction Motors Used by Electric Vehicles. *Trans. China Electro. Soc.* **2019**, *34*, 1488–1495.
23. Ye, D.L.; Li, J.; Qu, R.H.; Jiang, D.; Xiao, L.F.; Lu, Y. Variable Switching Sequence PWM Strategy of Dual Three-Phase Machine Drive for High-Frequency Current Harmonic Suppression. *IEEE Trans. Power Electron.* **2020**, *35*, 4984–4995. [CrossRef]
24. Huang, Y.; Xu, Y.; Zhang, W.; Zou, J. Hybrid RPWM Technique Based on Modified SVPWM to Reduce the PWM Acoustic Noise. *IEEE Trans. Power Electron.* **2019**, *34*, 5667–5674. [CrossRef]
25. Zhang, G.Y.; Wu, Z.Y.; Jiang, T. Research on Combination of Random Zero-Vector PWM Controller. *Electr. Drive Locomot.* **2018**, *1*, 57–61+67.
26. Bu, F.F.; Pu, T.; Huang, W.; Zhu, L. Performance and Evaluation of Five-Phase Dual Random SVPWM Strategy With Optimized Probability Density Function. *IEEE Trans. Ind. Electron.* **2019**, *66*, 3323–3332. [CrossRef]

27. Clerc, M.; Kennedy, J. The particle swarm—explosion, stability, and convergence in a multidimensional complex space. *IEEE Trans. Evol. Comput.* **2002**, *1*, 58–73. [CrossRef]
28. Yuan, X.; Chen, C.; Jiang, M.; Yuan, Y. Prediction interval of wind power using parameter optimized Beta distribution based LSTM model. *Appl. Soft Comput.* **2019**, *82*, 105550. [CrossRef]

electronics

MDPI

Article

Stochastic Operation of BESS and MVDC Link in Distribution Networks Under Uncertainty

Changhee Han [1], Sungyoon Song [2,*] and Jaehyeong Lee [3,*]

[1] Department of Electrical Engineering, Gyeongsang National University, Jinju-si 52828, Gyeongsangnam-do, Republic of Korea; chhan@gnu.ac.kr
[2] Department of Energy and Electrical Engineering, Tech University of Korea, Siheung-si 15117, Gyonggi-do, Republic of Korea
[3] Department of Electrical Engineering, Mokpo National University, Muan-gun 58554, Jeollanam-do, Republic of Korea
* Correspondence: songsy@tukorea.ac.kr (S.S.); jaehyeong@mnu.ac.kr (J.L.)

Abstract

This study introduces a stochastic optimization framework designed to effectively manage power flows in flexible medium-voltage DC (MVDC) link systems within distribution networks (DNs). The proposed approach operates in coordination with a battery energy storage system (BESS) to enhance the overall efficiency and reliability of the power distribution. Given the inherent uncertain characteristics associated with forecasting errors in photovoltaic (PV) generation and load demand, the study employs a distributionally robust chance-constrained optimization technique to mitigate the potential operational risks. To achieve a cooperative and optimized control strategy for MVDC link systems and BESS, the proposed method incorporates a stochastic relaxation of the reliability constraints on bus voltages. By strategically adjusting the conservativeness of these constraints, the proposed framework seeks to maximize the cost-effectiveness of DN operations. The numerical simulations demonstrate that relaxing the strict reliability constraints enables the distribution system operator to optimize the electricity imports more economically, thereby improving the overall financial performance while maintaining system reliability. Through case studies, we showed that the proposed method improves the operational cost by up to 44.7% while maintaining 96.83% bus voltage reliability under PV and load power output uncertainty.

Keywords: MVDC link; battery energy storage; chance-constrained optimization; distribution network

1. Introduction

As the international drive toward carbon neutrality proliferates, the integration of distributed generators (DGs) based on renewable energy sources (RESs) into distribution networks (DNs) has been increasing steadily [1–7]. This trend introduces significant challenges, due to the inherent intermittency and variability of RESs. Such fluctuations can lead to unpredictable negative impacts on future DNs, including voltage instability and load imbalances, as documented in prior studies [2,8–11]. In response to these challenges, the present study investigates the utilization of flexible medium-voltage DC (MVDC) link facilities to mitigate the voltage instability and load imbalances that occur across different DNs [12–17]. Specifically, these studies implement back-to-back voltage source converter systems (BTB-VSCs) within medium-voltage DNs as an interconnection

https://doi.org/10.3390/electronics14132737

mechanism to flexibly link different DNs [18–22]. The previous literature has employed various terminologies—such as Soft-Open Point, Power Flow Controller, and MVDC link—to describe the BTB-VSC configuration in DNs; yet, these terms fundamentally refer to the same system architecture.

Recognizing the limitation of the DC system, which is primarily limited to controlling only power flows between DNs, the study proposes a coordinated operational strategy that integrates battery energy storage systems (BESSs) into the network infrastructure. By leveraging the dual capabilities of BESSs for charging and discharging, the proposed strategy aims not only to correct the power imbalances across multiple DNs but also to enhance the overall operational performance for the distribution system operator (DSO). Such improvements are expected to enhance the economic performance by increasing the energy sales revenue and reducing the power importing cost.

Although prior research has advanced in improving DN operational efficiency through DC link integration, these studies have exhibited notable limitations; for example, the approaches in [23,24] employed robust optimization techniques that focused on worst-case scenarios arising from RES forecast errors, thereby achieving a minimal rate of constraint violations. However, such conservative methodologies may unexpectedly degrade operational efficiency, particularly in minimizing operational costs.

Alternatively, the research presented in [25,26] adopted stochastic optimization frameworks that modeled uncertainties using probabilistic functions, which allowed for more flexible DN operations. Despite this flexibility, the requirement to predefine the mathematical form of the probability distribution limits the practical applicability of these models.

To overcome these shortcomings, the proposed study employs a distributionally robust chance constrained optimization (DRCCO) method. This approach is designed to maintain robust performance in stochastic optimization without necessitating prior knowledge of the exact probability distributions of uncertainty variables.

Unlike traditional robust optimization, which is overly conservative, or scenario-based methods that require predefined distributions and a large number of samples, the DRCCO framework offers a flexible trade-off between reliability and cost without relying on strict distributional assumptions. Additionally, compared to the scenario-based approach in [27], which uses metaheuristic algorithms and normal distribution assumptions, our method is solver-friendly, distribution-free, and better suited for practical MVDC-based DN operations.

Within this framework, the optimal power flow control between DNs is achieved via the MVDC link, while the coordinated integration of BESSs further increases the operational benefits for the DSO. The validity and effectiveness of this integrated strategy are demonstrated through a case study based on the modified IEEE 33-bus test system. Additionally, the study conducts an in-depth analysis of several key parameters that influence the trade-off performance of the DRCCO model, thereby providing comprehensive insights into the optimization process and its practical implications for future DNs. Compared to previous research, the main contributions of this study are summarized as follows:

- In this work, we propose a day-ahead optimal scheduling strategy for MVDC links that aims to reduce the energy procurement costs for DSOs by coordinating with BESS under forecast uncertainties. The proposed method incorporates the DRCCO framework to optimize power allocation and ensure reliable operation across interconnected distribution networks.
- Furthermore, we conduct detailed case studies to investigate the impact of critical DRCCO parameters—namely, the ambiguity set radius and the confidence level—on the operational cost, voltage reliability, and energy loss. These analyses offer valuable insights for DSOs seeking to balance economic efficiency and system robustness.

The remainder of this paper is structured as follows. Section 2 introduces the MVDC link and presents the mathematical formulation of the proposed DRCCO problem. Section 3 discusses the simulation results and performance evaluation. Finally, Section 4 concludes the paper by summarizing the key findings.

2. Problem Formulation

2.1. Flexible MVDC Link System

As illustrated in Figure 1, the system under consideration consists of a single DC link shared by two VSCs, which are interconnected in a BTB configuration through a relatively short DC transmission line. Each VSC is responsible for establishing a connection with a distinct DN, thereby enabling controlled power exchange between the networks. The steady-state operational characteristics of this system can be mathematically formulated as follows [28,29].

$$\sum_{k=1}^{2}\left(P_{conv.t}^{k}+P_{conv.loss.t}^{k}\right)=0\,, \tag{1}$$

$$P_{conv.loss.t}^{k}=C_{conv.loss}\sqrt{\left(P_{conv.t}^{k}\right)^{2}+\left(Q_{conv.t}^{k}\right)^{2}}, \tag{2}$$

$$\sqrt{\left(P_{conv.t}^{k}\right)^{2}+\left(Q_{conv.t}^{k}\right)^{2}}\leq S_{conv}, \tag{3}$$

where, $P_{conv.t}^{k}$ and $P_{conv.loss.t}^{k}$ represent the active power output of k-th converter in the MVDC link. $C_{conv.loss}$ is the loss coefficient of the converter, which is generally set to 0.02. $Q_{conv.t}^{k}$ and S_{conv} are the reactive power output and rated capacity of k-th converter, respectively.

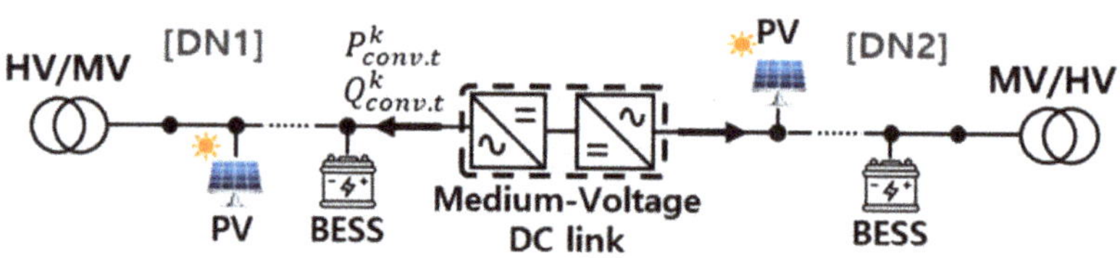

Figure 1. Configuration of interconnected DNs with MVDC link.

In this formulation, Equation (1) represents the fundamental principle of power distribution within the MVDC link, ensuring that the total power exchanged through the DC link is balanced. The power losses within the converters are captured in (2). Additionally, Equation (3) represents constraints on the maximum power output of the converters, thereby ensuring operational feasibility within predefined technical limits.

To facilitate computational efficiency and tractability, the nonlinear constraint in (2) can be relaxed into a second-order cone inequality, as shown in (4). The validity of this relaxation has been extensively studied in the prior research [23,24], demonstrating its effectiveness in accurately approximating converter losses while preserving the solution feasibility.

$$\sqrt{\left(P_{conv.t}^{k}\right)^{2}+\left(Q_{conv.t}^{k}\right)^{2}}\leq\frac{P_{conv.loss.t}^{k}}{C_{conv.loss}} \tag{4}$$

2.2. Battery Energy Storage System Model

The operational model of a BESS within DNs is formulated to account for its ability to perform both active power charging and discharging while enabling independent control of active and reactive power outputs. The mathematical representation of this model is structured as follows [30,31].

$$\sqrt{\left(P^k_{bess.t}\right)^2 + \left(Q^k_{bess.t}\right)^2} \leq S_{bess}, \tag{5}$$

$$P^k_{bess.loss.t} = C_{bess.loss}\sqrt{\left(P^k_{bess.t}\right)^2 + \left(Q^k_{bess.t}\right)^2}, \tag{6}$$

$$SOC^k_{(t+1)} = SOC^k_{(t)} - \left(P^k_{bess.t} + P^k_{bess.loss.t}\right)\Delta t, \tag{7}$$

$$SOC_{min} \leq SOC^k_{(t)} \leq SOC_{max}, \tag{8}$$

where, Equation (5) imposes constraints on the maximum output capacity of the BESS. $P^k_{bess.t}$ and $Q^k_{bess.t}$ represent the active and reactive power outputs, respectively, of the BESS installed in the k-th DN, with a rated power capacity of S_{bess}. The power losses associated with the DC/DC converter in BESS ($P^k_{bess.loss.t}$) are characterized in (6), which account for the conversion inefficiencies using the device loss coefficient ($C_{bess.loss}$). To effectively manage the energy stored within the BESS, Equation (7) establishes constraints on the state-of-charge (SOC), meaning that it operates within an acceptable range to maintain the battery health. Lastly, Equation (8) specifies the allowable SOC range by setting upper and lower bounds, thereby preventing the overcharging or excessive depletion of the BESS.

Similar to the relaxation of the MVDC link losses, Equation (6) can also be approximated to a second-order cone constraint as shown in (9).

$$\sqrt{\left(P^k_{bess.t}\right)^2 + \left(Q^k_{bess.t}\right)^2} \leq \frac{P^k_{bess.loss.t}}{C_{bess.loss}} \tag{9}$$

2.3. Load Flow Calculation Model

In this study, the power flow constraints are modeled using the DistFlow method, which provides an accurate representation of power flows in radial DNs [32]. This method is particularly effective for capturing the relationship between bus voltages, power injections, and line flows in distribution systems. The mathematical formulation of the DistFlow-based power flow constraints is presented as follows [32].

$$U^k_{i.t} = U^k_{j.t} - 2\left(r^k_{ij}P^k_{ij.t} + x^k_{ij}Q^k_{ij.t}\right) + \left(\left(r^k_{ij}\right)^2 + \left(x^k_{ij}\right)^2\right)l^k_{ij.t}, \tag{10}$$

$$\sum_{i:i\to j} P^k_{ij.t} - \sum_{i:i\to j} r^k_{ij}l^k_{ij.t} - \sum_{h:j\to h} P^k_{jh.t} = P^k_{load.j.t} - P^k_{pv.j.t} - P^k_{conv.t} - P^k_{bess.j.t}, \tag{11}$$

$$\sum_{i:i\to j} Q^k_{ij.t} - \sum_{i:i\to j} x^k_{ij}l^k_{ij.t} - \sum_{h:j\to h} Q^k_{jh.t} = Q^k_{load.j.t} - Q^k_{pv.j.t} - Q^k_{conv.t} - Q^k_{bess.j.t}, \tag{12}$$

$$\frac{\left(P^k_{ij.t}\right)^2 + \left(Q^k_{ij.t}\right)^2}{U^k_{i.t}} \leq l^k_{ij.t}, \tag{13}$$

where, the squared bus voltage and line current are defined as $U^k_{i.t} = \left(V^k_{i.t}\right)^2$ and $l^k_{i.t} = \left(I^k_{ij.t}\right)^2$. r^k_{ij} and x^k_{ij} denote the resistance and reactance magnitudes of branch (i, j) in k-th DN, respectively.$P^k_{ij.t}$ and $Q^k_{ij.t}$ represent the active and reactive power flows into branch (i, j) at time t in k-th DN, respectively. $P^k_{pv.j.t}$ and $Q^k_{pv.j.t}$ mean the active/reactive power output of the PV generators, $P^k_{load.j.t}$ and $Q^k_{load.j.t}$ also represent the load demands, respectively.

2.4. Uncertainty Variable Model

To effectively capture uncertainties from forecast error when predicting load demand and PV power generation outputs, this study models them as a single aggregated uncertainty variable for each DN connected to the MVDC link. The mathematical representation of this uncertainty modeling is provided as follows.

$$\tilde{P}^k_{load.j.t} = P^k_{load.j.t}\left(1 + \xi^k_{load.t}\right),\ \forall j \in \Omega^k_{load}, \tag{14}$$

$$\tilde{P}^k_{pv.j.t} = P^k_{pv.j.t}\left(1 + \xi^k_{pv.t}\right),\ \forall j \in \Omega^k_{pv}, \tag{15}$$

where superscript $(\tilde{\cdot})$ denotes the value after the error has been applied to the predicted value of the variable. $\xi^k_{pv.t}$ denotes the uncertainty variables for PV output forecasts. Specifically, it quantifies the proportional deviation of the actual PV output from the forecasted value at time t, reflecting PV generation uncertainty. $\xi^k_{load.t}$ represents the uncertainty variables for load demand forecasts. This also represents the proportional deviation of the actual load demand from the forecasted value at time t, capturing the load forecast uncertainty. Both coefficients range from 0 to 1. $P^k_{pv.j.t}$ and $P^k_{load.j.t}$ represent the day-ahead forecasted values of PV generation and load demand for bus j at time t in the k-th DN. These forecast values are obtained based on historical data and predictive models.

Finally, the bus voltage, which reflects the forecast errors of PV generation and load output, as well as the effects of the outputs from the MVDC link and BESS, can be expressed using sensitivity coefficients as follows.

$$\tilde{V}^k_{i.t} = V^k_{i.t} + \Delta\tilde{V}^k_{load.i.t} + \Delta\tilde{V}^k_{pv.i.t} + \Delta V^k_{conv.i.t} + \Delta V^k_{bess.i.t} \tag{16}$$

$$\Delta\tilde{V}^k_{load.i.t} = \sum_{j\in\Omega^k_{load}} \left\{ \frac{\partial V^k_i}{\partial P^k_j} P^k_{load.j.t} + \frac{\partial V^k_i}{\partial Q^k_j} Q^k_{load.j.t} \right\} \left(\xi^k_{load.t}\right) \tag{17}$$

$$\Delta\tilde{V}^k_{pv.i.t} = \sum_{j\in\Omega^k_{pv}} \left\{ \frac{\partial V^k_i}{\partial P^k_j} P^k_{pv.j.t} + \frac{\partial V^k_i}{\partial Q^k_j} Q^k_{pv.j.t} \right\} \left(\xi^k_{pv.t}\right) \tag{18}$$

$$\Delta V^k_{conv.i.t} = \frac{\partial V^k_i}{\partial P^k_j} P^k_{conv.t} + \frac{\partial V^k_i}{\partial Q^k_j} Q^k_{conv.t} \tag{19}$$

$$\Delta V^k_{bess.i.t} = \frac{\partial V^k_i}{\partial P^k_j} P^k_{bess.t} + \frac{\partial V^k_i}{\partial Q^k_j} Q^k_{bess.t} \tag{20}$$

While there are various previous studies related to the calculation methods for voltage-to-power sensitivity, this study adopts the methodology in [33] due to its relatively fast computation speed.

While the sensitivity coefficients were calculated using the method in [33] for computational efficiency, we acknowledge that this approach involves linear approximations. To ensure that the DRCCO reliability guarantees were met, we performed a Monte Carlo-based scenario validation using 3000 additional forecast error samples and full nonlinear power flow calculations, as described in Section 3.

2.5. Distributionally Robust Chance-Constrained Optimization Model

Chance-constrained optimization is an advanced optimization technique that ensures constraints are satisfied with a predefined confidence level in problems involving uncertainty. This approach formulates the optimization problem such that the probability of constraint violations does not exceed a specified threshold, thereby enhancing the reliability of decision-making under uncertainty.

Traditional studies addressing chance-constrained optimization typically assume that the probability distribution functions are precisely known. However, in practical applications, it is often difficult to assert with confidence that uncertainty factors—such as forecast errors in PV generation and load demand—adhere to specific probability distributions, such as Gaussian distribution. This uncertainty in distribution modeling necessitates the use of a methodology to quantify the distance between different probability distributions.

Due to these limitations, the concept of DRCCO has been proposed [34]. As illustrated in Figure 2, an ambiguity set is constructed, which encompasses various potential probability distributions of the uncertain variables. This ambiguity set is typically defined on the empirical distribution function derived from sample data of the uncertainty. It is formulated to include all distributions within a certain radius around the empirical distribution, thereby considering the true distribution (but unknown) in a realistic manner. Among various measures of distributional distance, this study adopts the Wasserstein distance [34] for its robustness and flexibility in addressing distributional uncertainty.

$$d_W(\mathbb{Q}_1, \mathbb{Q}_2) := \sup_{f \in \mathcal{L}} \left\{ \int_{\Xi} f(\xi)\mathbb{Q}_1(d\xi) - \int_{\Xi} f(\xi)\mathbb{Q}_2(d\xi) \right\}, \tag{21}$$

$$\mathcal{B}_\varepsilon\left(\hat{\mathbb{P}}_N\right) := \left\{\mathbb{P} \in \mathcal{M}(\Xi) : d_W\left(\mathbb{P}, \hat{\mathbb{P}}_N\right) \le \varepsilon\right\}, \tag{22}$$

$$\inf_{\mathbb{P} \in \mathcal{B}_\varepsilon(\hat{\mathbb{P}}_N)} \mathbb{P}\left(V_{min} \le \overset{\sim k}{V}_{i.t} \le V_{max}\right) \ge 1 - \alpha_V, \tag{23}$$

where Equation (21) defines the Wasserstein distance used to measure the difference between two probability distributions $(\mathbb{Q}_1, \mathbb{Q}_2)$, while Equation (22) specifies that all distributions within the ambiguity set must lie within a certain radius (ε) from a reference distribution $(\hat{\mathbb{P}}_N)$. V_{min} and V_{max} denote the allowable bus voltage limits, while $(1 - \alpha_V)$ represents the minimum confidence level for constraint satisfaction. For instance, $\alpha_V = 0.05$ limits the bus voltage violation probability to 5% under PV and load demand forecast errors. This constraint ensures that the true distribution lies within a Wasserstein ball of radius ε centered at the empirical distribution, enabling the model to protect against sampling errors and distributional shifts while maintaining tractability.

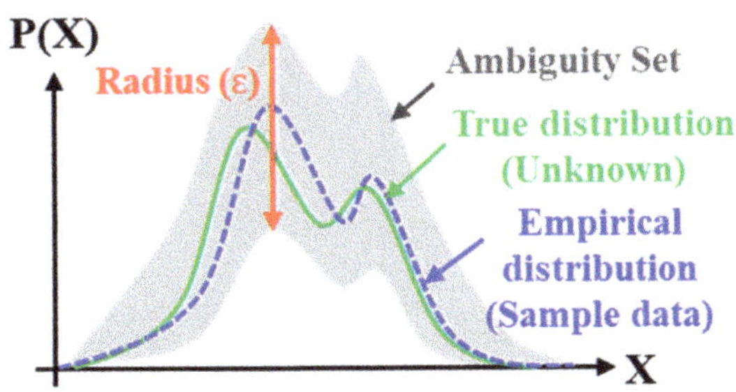

Figure 2. Illustration of ambiguity set in DRCCO framework.

The choice of the Wasserstein distance for defining the ambiguity set was motivated by several factors [34–38]. First, the Wasserstein metric offers greater modeling flexibility in scenarios with limited sample sizes, as it does not require the empirical distribution to be absolutely continuous with respect to the true distribution, unlike other measure such as the Kullback–Leibler divergence [38]. Second, it captures both the shape and support of the distributions, allowing for better characterization of the distributional shifts that may occur under extreme PV and load forecast errors. Finally, the DRCCO problem formulated with the Wasserstein distance can be converted into a tractable convex optimization problem, enabling an efficient solution using standard solvers. Although this study did not perform a direct comparison with other distance metrics, the sensitivity analysis as shown in the following tables and figures in Section 3 illustrates how varying the radius of the Wasserstein ambiguity set impacts the operational cost and voltage reliability.

In (23), both lower and upper constraints are included. Therefore, the inequalities for the left-hand side and right-hand side can be separated and expressed as (24) and (25). In the case of voltage, it can be expressed as a linear combination of the active/reactive power output variables of the MVDC link and BESS, as well as uncertainty variables, similar to (16). Thus, it can be represented by the following Equations:

$$\inf_{\mathbb{P}\in\mathcal{B}_\varepsilon(\hat{\mathbb{P}}_N)} \mathbb{P}(\ell_{V.t}(x,\xi) \le 0) \ge 1-\alpha_V\,, \tag{24}$$

$$\ell_{V.t}(x,\xi) = \max_{k\in N_d, i\in\Omega_b^k}\left\{\left(\tilde{V}^k_{i.t} - V_{max}\right), \left(V_{min} - \tilde{V}^k_{i.t}\right)\right\}, \tag{25}$$

$$\ell_{V.t}(x,\xi) = \max_{n\in N_p}(\boldsymbol{a}_{n.t}\boldsymbol{X}_{conv.t} + \boldsymbol{b}_{n.t}\boldsymbol{\xi}_t + \boldsymbol{c}_{n.t}), \tag{26}$$

where the vector $\boldsymbol{X}_{conv.t} = \left[P^k_{conv.t}, Q^k_{conv.t}, P^k_{bess.t}, Q^k_{bess.t}\right]$ represents the set of decision variables, which includes the active and reactive power outputs of both the MVDC link and the BESS at time t. The vector $\boldsymbol{a}_{n.t}$ denotes the voltage sensitivities of bus voltages in the DN, as expressed in (19) and (20). The vector $\boldsymbol{b}_{n.t}$ consists of coefficients that correspond to the forecast errors in PV generation and load demand, as detailed in (17) and (18). These coefficients account for the uncertainty associated with the forecasting of generation and demand. Finally, $\boldsymbol{c}_{n.t}$ represents the differences between the predicted voltage and the maximum/minimum allowable voltage limits. Specifically, it includes the terms $(\left(V^k_{i.t} - V_{max}\right), \left(V_{min} - V^k_{i.t}\right))$, which quantify the deviation from the upper and lower voltage constraints, respectively.

The implementation of the DRCCO in (24) using standard optimization solvers can be facilitated by applying the results from earlier research [34], which enable the problem to be reformulated in its dual form as follows:

$$\lambda_t\varepsilon + \frac{1}{N(\Omega_s)}\sum\nolimits_{k\in\Omega_s}\delta_{t.k} \le \alpha_V\theta_t\,, \tag{27}$$

$$\left(\boldsymbol{a}_{n.t}\boldsymbol{X}_{conv.t} + \boldsymbol{b}_{n.t}\bar{\boldsymbol{\xi}}_t + \boldsymbol{c}_{n.t}\right) + \boldsymbol{\rho}^T_{n.i.t}\left(\boldsymbol{d} - \boldsymbol{C}\bar{\boldsymbol{\xi}}_t\right) + \boldsymbol{\theta}_t \le \delta_{t.k}, \tag{28}$$

$$\left\|\boldsymbol{C}^T\boldsymbol{\rho}^T_{n.i.t} - \boldsymbol{b}_{n.t}\right\|_\infty \le \lambda_t\,, \tag{29}$$

$$\lambda_t \ge 0,\ \delta_{t.k} \ge 0,\ \boldsymbol{\rho}^T_{n.i.t} \ge 0, \tag{30}$$

where the vector $\boldsymbol{X}_{dual.t} = [\boldsymbol{\rho}_{n.i.t}, \delta_{t.k}, \boldsymbol{\theta}_t, \lambda_t]$ denotes the decision variables in the dual problem.

It is worth noting that the relaxation of bus voltage reliability constraints within the DRCCO framework is bounded by a predefined confidence level (e.g., 95%), which quantitatively limits the risk of constraint violation under uncertainty. This does not imply a relaxation of the absolute voltage limits specified in standards such as IEEE 1547 or IEC 61,970 but rather allows for probabilistic operation under high-variability conditions. This approach is consistent with emerging grid codes that allow limited statistically controlled deviations in the presence of renewable generation. Moreover, if necessary, reactive power control through BESS and MVDC converters can provide dynamic voltage support, mitigating any adverse impacts on long-term system stability.

2.6. Proposed Optimization Model

Through the proposed coordinated operation of the MVDC link and BESS, the DSO can focus on reducing operational costs within the DN or maximizing profits derived from power sales to the upstream grid. To achieve this, the objective function is formulated as the cumulative value of electricity procurement costs over the course of a day for the DNs that are interconnected through the MVDC link. Since the operational objective of the DSO is to minimize the expected value of the electricity procurement costs across all DNs interconnected via the MVDC link and BESS, the expected value of the sum of net loads for the two DNs can be formulated as (31) and (32).

$$\overset{\sim k}{P}_{nl.t} = \sum_{j\in\Omega_b^k}\left(\overset{\sim k}{P}_{load.j.t} - \overset{\sim k}{P}_{pv.j.t}\right) + \sum_{(i,j)\in\Omega_l^k}\left(r_{ij}^k \overset{\sim k}{l}_{ij}\right) - P_{conv.t}^k \tag{31}$$

$$\mathbb{E}\left[\sum_{k=1}^{2}\overset{\sim k}{P}_{nl.t}\right] = \sum_{k=1}^{2}\left\{\sum_{j\in\Omega_b^k}\left(P_{load.j.t}^k - P_{pv.j.t}^k - P_{bess.j.t}^k\right) + \sum_{(i,j)\in\Omega_b^k}\left(r_{ij}^k l_{ij}^k\right) + P_{conv.loss.t}^k\right\} \tag{32}$$

This formulation assumes that the forecast errors for the load demand and PV power generation, as represented in (14) and (15), have an expected value of zero. Consequently, this implies that the expected value of the line power flows, which are influenced by these uncertainties, is also zero. Furthermore, it is also important to note that the sum of the outputs in the MVDC link, along with the converter losses, is zero, as described in (1). This condition is crucial for ensuring power conservation within the MVDC link and simplifies the formulation of the optimization problem by removing any power imbalances in the interconnected DNs.

This study assumes a scenario where two distinct DNs are interconnected through the MVDC link. Based on this assumption, an optimal operation problem is formulated with the objective of minimizing the total electricity procurement cost across both DNs. The problem is structured as follows, considering the relevant operational constraints and the interactions between the MVDC link and BESS in DNs.

$$\min_{X_{conv}} \sum_{t=1h}^{T} C_{s.t}\left\{\sum_{(i,j)\in\Omega_b^k}\left(r_{ij}^k l_{ij}^k\right) + P_{conv.loss.t}^k - P_{bess.t}^k\right\}\Delta t, \quad s.t.(1)\text{–}(32) \tag{33}$$

The visual representation of the optimization problem proposed in this study is shown in Figure 3. Depending on the user's operational intent, the ambiguity set is constructed by inputting two key parameters from the DRCCO models: the radius of the ambiguity set and the violation probability, along with sample data of the uncertainty variables.

By utilizing the predicted daily profiles of the PV generation and load, along with hourly electricity cost data, the daily operation model of the DN, the operation model of the MVDC link and BESS, and the objective function are formulated. These components are integrated to construct the final optimization model for the probabilistic operation of the MVDC link and BESS. This approach ensures the coordinative operation between the MVDC link and BESS considering the uncertainty of PV and load outputs in DNs.

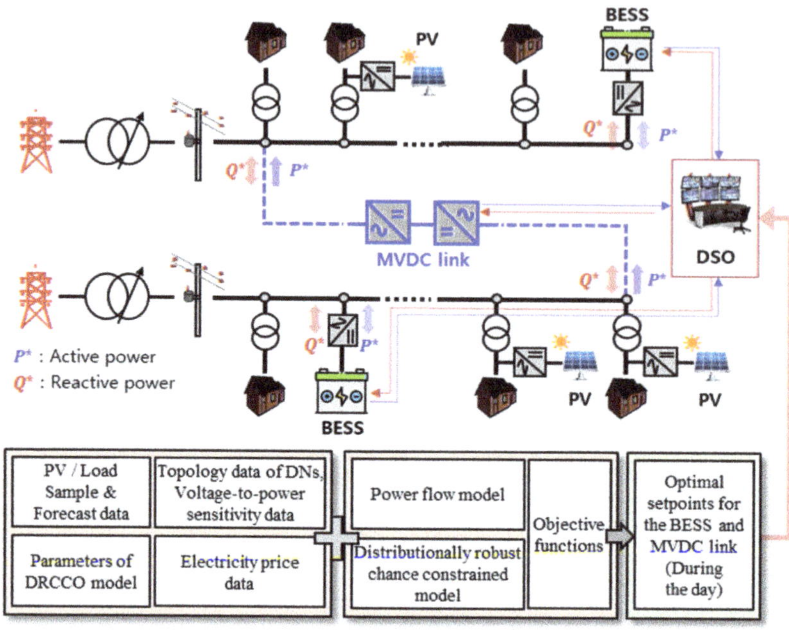

Figure 3. Architecture of proposed DRCCO framework to operate MVDC link and BESS.

Within the architecture of the proposed DRCCO framework in Figure 3, the power flow is managed in a coordinated manner between the interconnected DNs, the MVDC link, and the BESS units. The MVDC link facilitates bidirectional active power transfer between the two DNs, enabling the redistribution of excess generation or load support depending on real-time net load imbalances. Simultaneously, the BESS units in each DN perform active power charging and discharging operations to further mitigate forecast uncertainties and optimize the total procurement cost. During periods of high PV generation, surplus power can be stored in the BESS or exported to the neighboring DN via the MVDC link. Conversely, during peak load hours or when electricity prices are high, stored energy in the BESS can be discharged to reduce grid procurement costs.

The proposed DRCCO framework determines the optimal setpoints for MVDC link power flows and BESS outputs in each time step, ensuring that all operational constraints—such as voltage limits and power balance—are satisfied within the desired reliability level under forecast uncertainties. This integrated power flow coordination between the MVDC link and BESS plays a critical role in enhancing both economic efficiency and voltage stability across the interconnected DNs.

3. Case Studies

To validate the proposed approach in this study, a scenario was assumed where two IEEE 33-bus test systems are interconnected through the MVDC link, as illustrated in Figure 4. The detailed specifications of these test systems are summarized in Table 1. The IEEE 33-bus test system was chosen, as it is a widely used benchmark in DN studies, especially for MVDC link and BESS operation analysis [18–22].

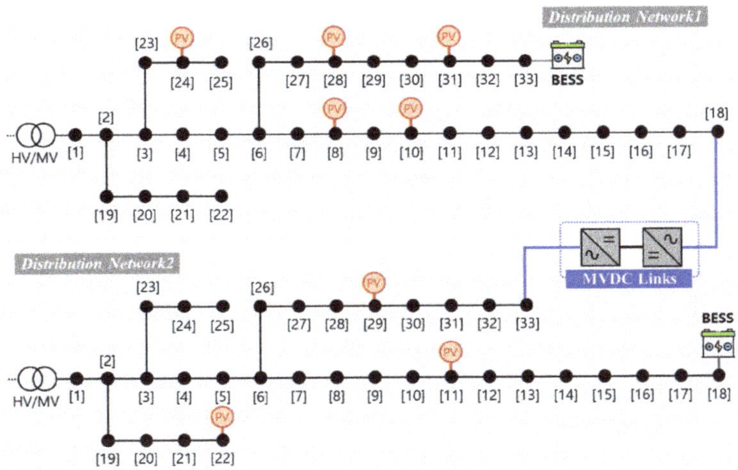

Figure 4. Test systems using modified IEEE-33 bus distribution networks.

Table 1. Detailed specification of test systems.

DN	PV Location	MVDC Link Location	BESS Location
1	8, 10, 24, 28, 33	18	33
2	11, 22, 29	33	18
DN	**PV capacity**	**MVDC Link Capacity**	**BESS Capacity**
1 and 2	700 kVA (Each)	1000 kVA	500 kVA/1500 kWh

As shown in Figure 4, DN1 has a higher PV penetration compared to DN2, and both the MVDC link and BESS units are placed at the end buses of each feeder to maximize their voltage regulation impact. The daily forecast profiles for PV generation, load demand, and hourly power purchase costs for each test system are shown in Figures 5 and 6 [39]. Figures 5 and 6 illustrate the daily load share and electricity price profiles, respectively. During the day, DN2 accounts for a higher share of the total load, while in the evening—when PV output drops to zero—DN1 becomes dominant. In addition, electricity import prices peak during 8:00–10:00 and 18:00–20:00.

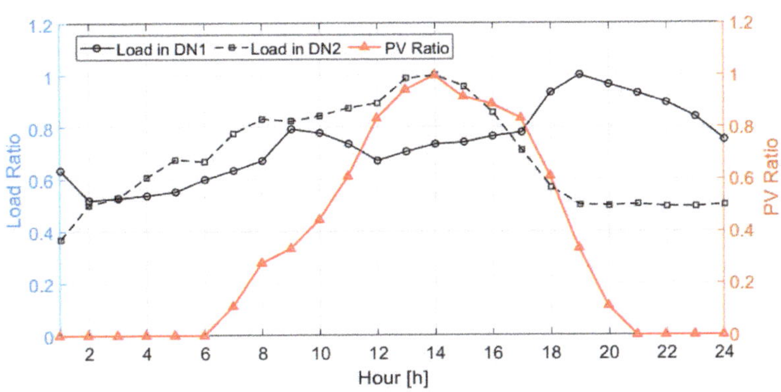

Figure 5. Hourly predicted PV output (**left**) and load demand profile (**right**).

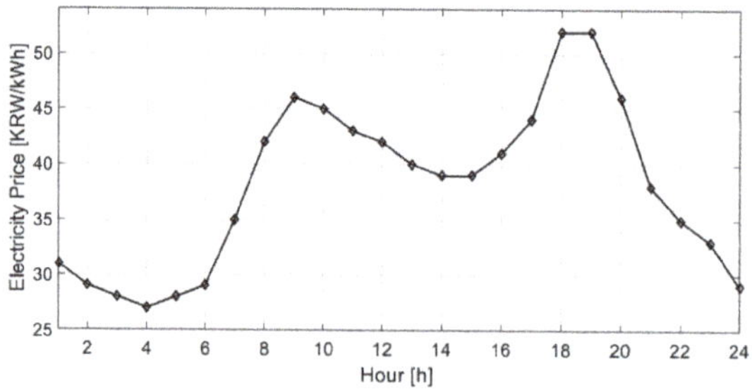

Figure 6. Hourly electricity price profile.

All simulations were performed using MATLAB R2022b. The optimization problems were implemented using the CVX modeling language and solved with the MOSEK solver.

Figures 7 and 8 present the results of determining the optimal power outputs of the MVDC link and the BESS in each DN over a day. As outlined in the optimal operation problem described in (33), this optimization was achieved by minimizing the operational costs while restricting the probability of constraint violations regarding bus voltage maintenance in the two interconnected DNs to a specified value. Using the predicted daily PV generation and load outputs, the proposed DRCCO model was applied to the interconnected DNs (DN1 and DN2). This approach ensured that the bus voltage confidence level remained within the desired probability (95%) through the optimal operation of the BESS and MVDC link.

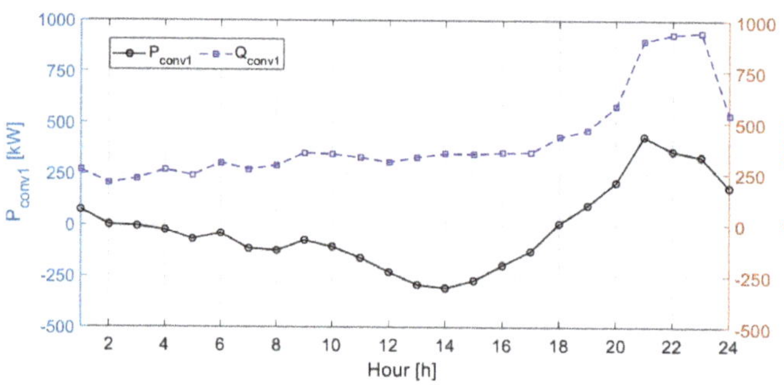

(**a**)

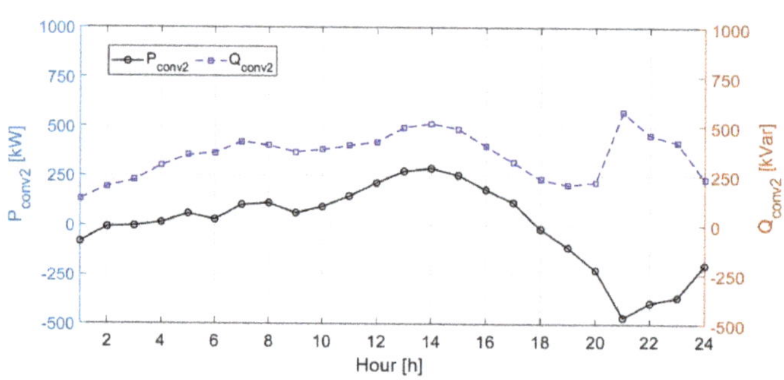

(**b**)

Figure 7. Hourly optimal power outputs of the MVDC link. ((**a**): Converter in DN1 and (**b**): converter in DN2).

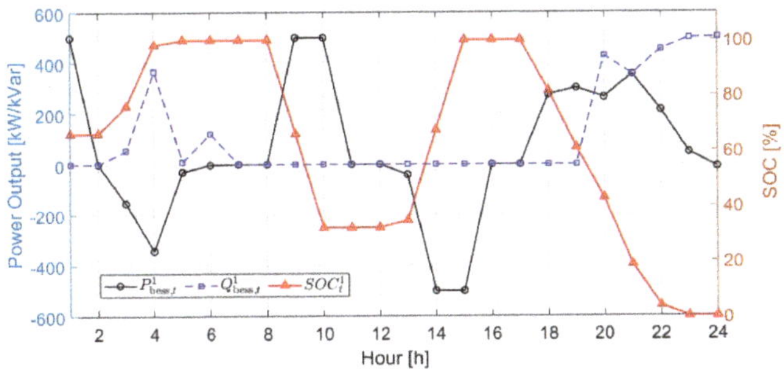

(a)

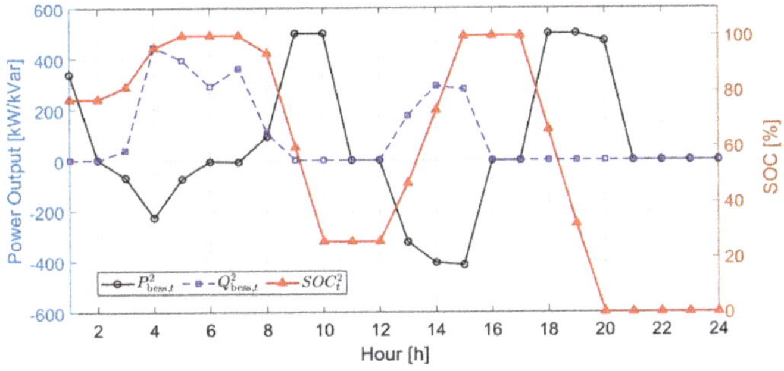

(b)

Figure 8. Hourly optimal power outputs of the BESS ((**a**): BESS in DN1 and (**b**): BESS in DN2).

To form the ambiguity set for uncertainty variables, such as forecast errors in PV generation and load demand, 30 samples were used in this study (N_S = 30).

As observed in Figure 7, the power imbalance caused by the differences in net load between the two DNs can be mitigated to some extent through the operation of the MVDC link, given the differences in their load profiles and PV capacities.

For example, during the 18–24 h period, when the load in DN1 is higher than in DN2, the MVDC link draws active power from DN2 and injects it into DN1, balancing the load between the two networks. When the net loads connected through the MVDC link are summed, the power injected through the MVDC link offsets each other in opposite directions, as shown in (32). This behavior ensures that the MVDC link operates in a way that reduces line losses across the interconnected DNs, thereby indirectly contributing to the minimization of operational costs.

In Figure 8, during periods of low electricity costs (4–6 AM, 2–4 PM), the BESS in each DN performs maximum charging to take advantage of the cheaper electricity. On the other hand, during high-cost periods, such as 8–11 AM and 6–8 PM, the BESS maximizes the discharge of stored power, effectively utilizing opportunities to reduce the overall procurement costs by discharging when electricity prices are higher. This strategy optimizes both the operational efficiency and the economic performance of the interconnected DNs. Furthermore, by controlling the reactive power output, the MVDC link and BESS in each DN play a key role in voltage stabilization, ensuring that the system can deliver maximum power output when necessary.

To further compare and verify the performance of the proposed method, four scenarios were considered:

- Scenario I: Baseline cases without the integration of the MVDC link and BESS.

- Scenario II: Incorporates the MVDC link and BESS, employing a deterministic optimization framework that neglects forecast uncertainties.
- Scenario III: Incorporates the MVDC link and BESS, employing a robust optimization that considers only the maximum and minimum values of forecast uncertainties.
- Scenario IV: Incorporates the MVDC link and BESS, employing the proposed DRCCO method to explicitly account for forecast uncertainties.

Table 2 presents a performance comparison across several quantitative indicators for each scenario. A comparison with Scenario I clearly shows that the coordinated operation of the BESS and MVDC link can transform the optimal costs of the two DNs into negative values, indicating that power sales generate profits. However, Scenario II, which applies a deterministic approach that neglects PV and load forecast uncertainties, shows the lowest optimal cost (USD −53.38/day) but suffers from severely degraded voltage reliability (75.17%). This low reliability arises because the deterministic method does not consider forecast errors, leading to frequent constraint violations in real operation. In contrast, Scenario III, based on robust optimization, guarantees 100% bus voltage reliability by considering worst-case forecast errors but at the expense of higher operational cost (USD −33.96/day). Notably, the proposed DRCCO approach (Scenario IV) effectively balances these two aspects, achieving a high voltage reliability of 96.83% while securing a 44.7% cost improvement compared to the robust method.

Table 2. Comparison of operational performance under different scenarios.

Scenarios	Average Energy Loss Per Day [kWh/day]	Optimal Cost [USD/day]	Bus Voltage Reliability [%]
I	3490.90	132.86	0
II	2237.10	−53.38	75.17
III	2403.20	−33.96	100
IV	2255.37	−49.14	96.83

Specifically, 3000 additional forecast error scenarios were generated for the predetermined operating points, power flow calculations were performed, and the number of scenarios where constraints were satisfied was counted to calculate the rate.

In contrast, Scenario III, which employs robust optimization, shows no constraint violations (100% reliability); however, its conservative nature leads to the least significant minimization of the objective function. Finally, in Scenario IV—using the proposed DRCCO—the bus voltage reliability is regulated to approximately 96.83%, resulting in an increase in the DSO's power sales revenue by up to $49.14 per day, which is an 44.7% improvement compared to the robust optimization approach.

Furthermore, Table 3 compares the performance based on the size of the bus voltage confidence level ($1 - \alpha_V$). As the set value for voltage violation increases, the DRCCO model allows for a broader range of constraints. This expansion leads to a lower bus voltage reliability (i.e., higher constraint violation rate); however, it also provides greater flexibility in the outputs of the MVDC link and BESS, thereby resulting in increased power sales revenue.

Lastly, Table 4 presents a performance comparison based on the radius of the ambiguity set employed in the proposed DRCCO problem. As the radius of the ambiguity set increases relative to the sample data, the candidate set for uncertainty variables is defined more broadly, which leads to more conservative solutions to the optimization problem. As presented in Table 4, an increase in the radius leads to higher average network energy losses and the increase in power sales revenue.

Table 3. Impact of voltage confidence level on operational performance.

Bus Voltage Confidence Level $(1-\alpha_V)$	Average Energy Loss Per Day [kWh/day]	Cost Improvement [%]	Bus Voltage Reliability [%]
95%	2255.37	44.7	96.83
90%	2251.02	47.1	95.40
85%	2247.87	49.8	92.40
80%	2244.49	51.2	90.03

Table 4. Impact of radius of ambiguity set on operational performance.

Radius of Ambiguity Set (ε)	Average Energy Loss Per Day [kWh/day]	Cost Improvement [%]	Bus Voltage Reliability [%]
0.001	2248.21	49.2	91.67
0.002	2250.54	47.1	94.73
0.003	2255.37	44.7	96.83
0.005	2270.79	38.7	98.63

Figure 9 visualizes this trade-off relationship between bus voltage reliability and cost improvement with respect to varying voltage confidence levels and radius of ambiguity sets. This highlights the inherent balance between economic efficiency and operational robustness in the proposed DRCCO-based dispatch strategy.

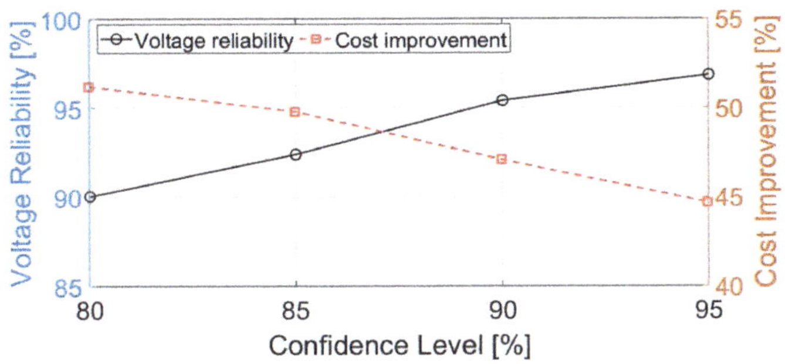

(**a**)

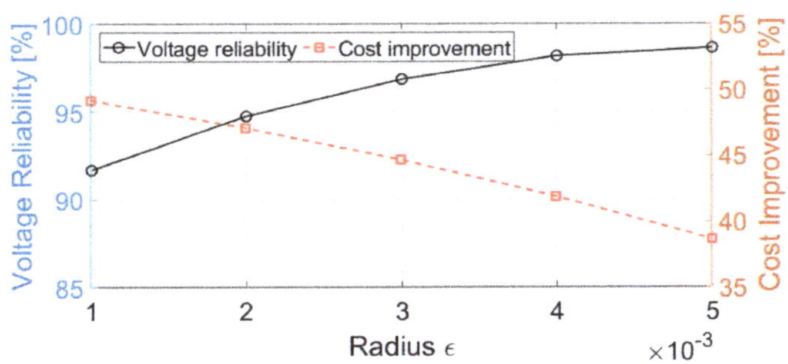

(**b**)

Figure 9. Comparison of system performance ((**a**): confidence level and (**b**): radius).

Consequently, the impact of two key parameters—namely, the confidence level $(1 - \alpha_V)$ and the radius of the ambiguity set (ε)—on system performance was evaluated using three primary metrics: operational cost (USD/day), bus voltage reliability (%), and cost improvement (%). Additionally, the average energy loss (kWh/day) was reported to capture the system efficiency. As shown in Tables 3 and 4, these indicators enable a quantitative understanding of the trade-off between economic benefits and reliability risks under different DRCCO settings.

The selection of the ambiguity set radius (ε) and the voltage confidence level $(1 - \alpha_V)$ was guided by a sensitivity analysis, as presented in Tables 3 and 4 and Figure 9. By evaluating the trade-off between operational cost and voltage reliability across different parameter settings, we identified that $\varepsilon = 0.003$ and $(1 - \alpha_V) = 95\%$ offered a reasonable balance. Further increases in parameter conservativeness beyond this point resulted in marginal reliability improvements but a sharp rise in operational cost.

Figure 10 and Table 5 present the impact of varying sample sizes on the cost improvement, voltage reliability, and computation time within the proposed DRCCO framework. As the sample size increases from 30 to 300, the voltage reliability improves steadily from 96.83% to 98.82%, while the cost improvement declines from 44.7% to 38.9%. Notably, the computation time increases sharply, from 53.05 [s] at 30 samples to 1550.37 [s] at 300 samples.

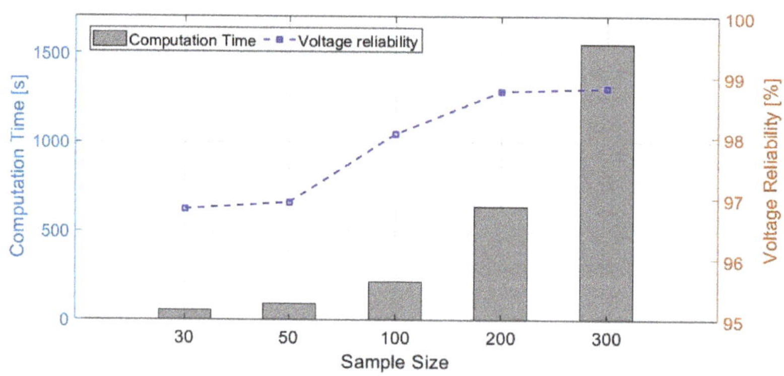

Figure 10. Comparison of system performance with the varying numbers of sample sizes.

Table 5. Comparison of performance indices under different sample sizes.

Sample Sizes	30	50	100	200	300
Cost Improvement [%]	44.7	42.8	41.0	39.7	38.9
Bus voltage reliability [%]	96.83	96.94	98.07	98.77	98.82
Computation time [s]	53.05	90.13	212.95	635.59	1550.37

Increasing the sample size improves voltage reliability but also leads to a substantial rise in computation time. For example, the reliability increases from 96.83% (30 samples) to 98.07% (100 samples), while the computation time grows from 53 to 213 s. Beyond 100 samples, the reliability gain becomes marginal, whereas the computation time increases sharply. Therefore, a sample size between 30 and 100 is recommended to balance the reliability and computational efficiency for day-ahead scheduling applications. These results highlight the trade-off between computational burden and performance accuracy, underscoring the importance of selecting an appropriate sample size that balances reliability, cost efficiency, and tractability.

Additionally, to evaluate the scalability of the proposed DRCCO framework for more complex DNs, we extended the case study to three DNs interconnected with three independent MVDC links. The detailed configuration of the system is illustrated in Figure 11. The data of DN1 (IEEE-33 bus system) are from [40], DN2 (IEEE-69 bus system) and DN3 (IEEE-85 bus system) are from [41] and [42], respectively. This setup enables enhanced power exchange flexibility among multiple DNs, each equipped with its own BESS.

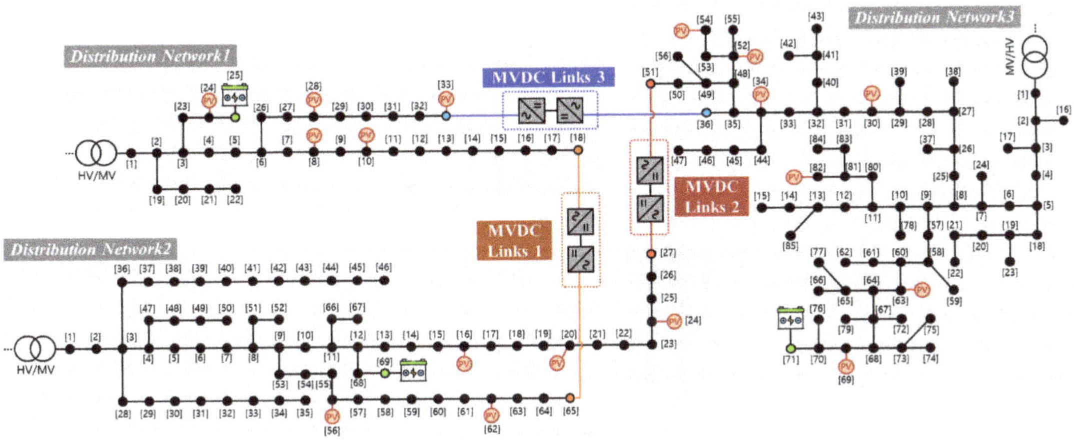

Figure 11. Configuration of three DNs interconnected via multiple MVDC links.

The detailed specifications of each test system are summarized in Table 6. As shown, the system includes three MVDC links (MVDC1, MVDC2, and MVDC3) connecting the DNs, with each DN equipped with its own BESS to enhance operational flexibility.

Table 6. Detailed specification of each test system in the multiple DNs.

DN	PV Location	MVDC Link Location MVDC1	 MVDC2	 MVDC3	BESS Location
1 (IEEE-33)	8, 10, 24, 28, 33	18	-	33	25
2 (IEEE-69)	16, 20, 24, 56, 62	65	27	36	69
3 (IEEE-85)	30, 34, 52, 54, 63, 69, 82	-	51	-	71
DN	**PV Capacity**	**MVDC Link Capacity**			**BESS Capacity**
1 and 2	700 kVA (Each)	1000 kVA			500 kVA/1500 kWh
3	500 kVA (Each)	1000 kVA			500 kVA/1500 kWh

Figures 12 and 13 illustrate the optimal power dispatch of MVDC links and BESS in the three DNs. These dispatch profiles reflect the hourly net load variations of each DN in Figure 14. During low net load periods in DN3 (e.g., 1:00–7:00), active power is transferred from DN3 to DN1 and DN2 through MVDC2 and MVDC3. Similarly, during midday hours (12:00–17:00), surplus power from DN1 and DN3 is delivered to DN2, while in the evening (18:00–24:00), DN2 and DN3 supply power to DN1 to meet its higher net load.

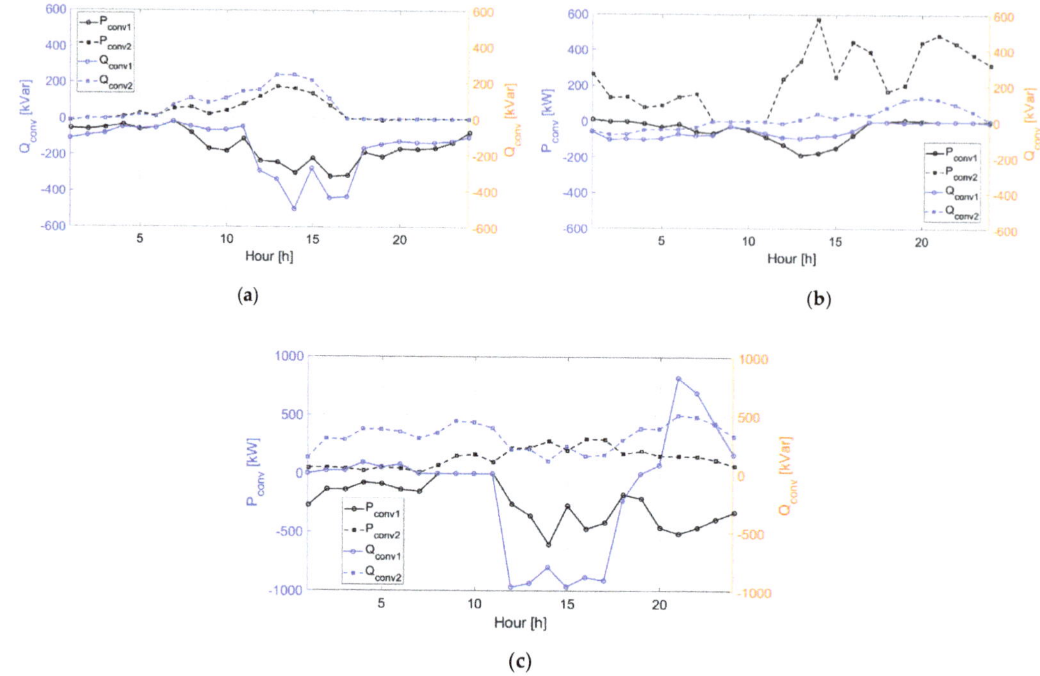

Figure 12. Hourly optimal power outputs of MVDC links in multiple DNs ((**a**): converters in DN1, (**b**): converters in DN2, and (**c**): converters in DN3).

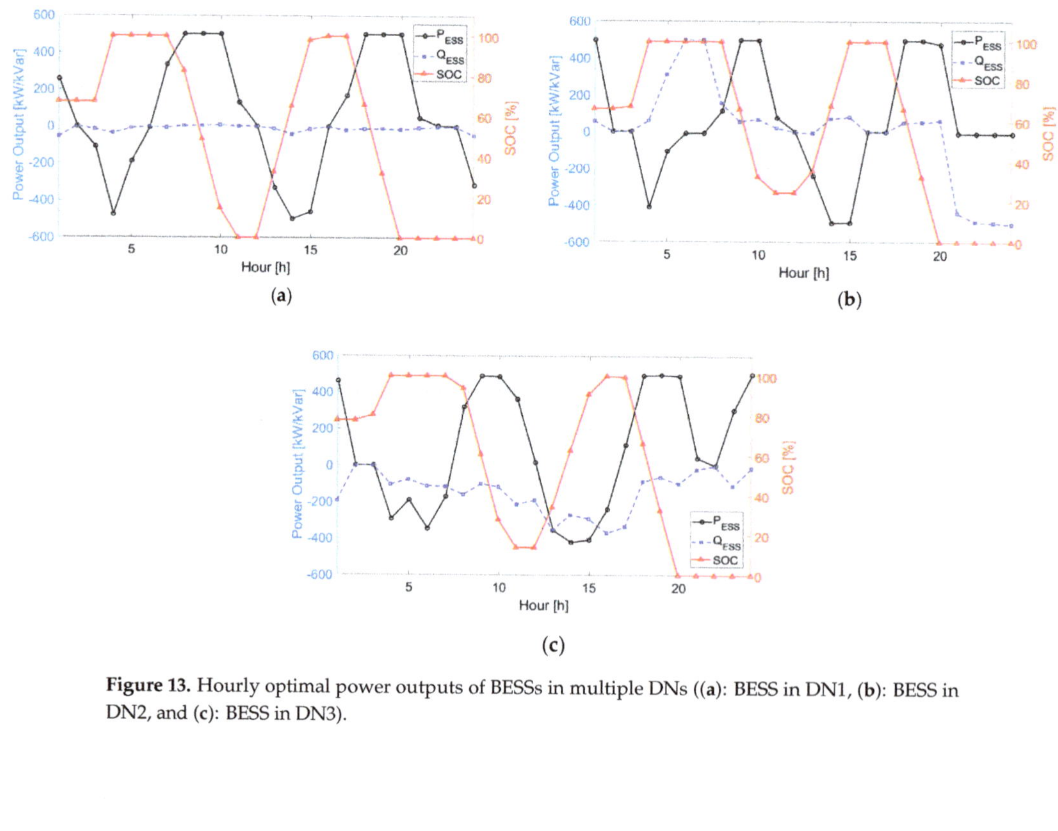

Figure 13. Hourly optimal power outputs of BESSs in multiple DNs ((**a**): BESS in DN1, (**b**): BESS in DN2, and (**c**): BESS in DN3).

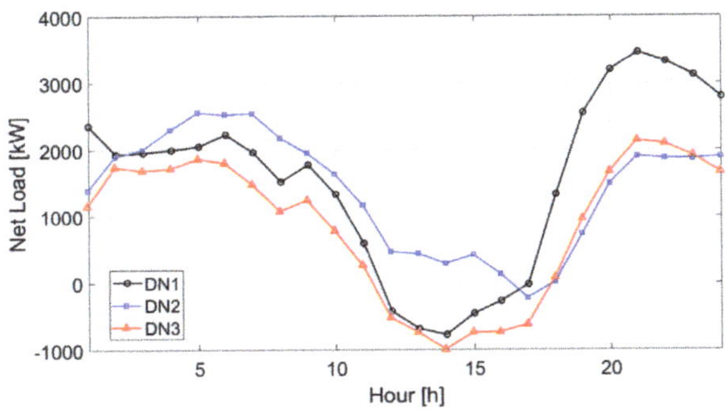

Figure 14. Hourly net load profiles in each DN.

In parallel, the BESS units perform charging during low-price periods and discharging during high-price hours. The coordinated operation of MVDC links and BESSs effectively balances the net load differences among the DNs while minimizing the operational costs under the proposed DRCCO framework.

Under the simulation using a three-DN interconnected test system, Table 7 summarizes the performance comparison for Scenarios II, III, and IV under different voltage confidence levels. Similar to the previous case study, Scenario II (deterministic approach) achieved the lowest cost but showed poor voltage reliability (25.4%). Scenario III (robust approach) ensured 100% reliability but with the highest cost. In contrast, Scenario IV (proposed DRCCO method) achieved a well-balanced trade-off, maintaining high voltage reliability (above 90%) while significantly reducing the operational costs compared to the robust approach. In addition, the calculation time for determining the day-ahead MVDC link set-points under Scenario IV was 146.24 s, demonstrating the computational efficiency and practical applicability of the proposed method from the perspective of computational burden.

Table 7. Detailed specification of each test system.

Scenarios	Bus Voltage Confidence Level $(1-\alpha_V)$	Optimal Cost [USD/Day]	Cost Improvement [%]	Bus Voltage Reliability [%]
II	-	325.19	35.17	25.4
III	100%	501.64	-	100
IV	95%	403.12	19.64	98.86
	90%	381.11	24.03	94.60
	85%	371.40	25.96	90.30
	80%	365.52	27.13	84.50

Furthermore, the current study focuses solely on operational cost minimization for day-ahead scheduling, assuming that the MVDC link and BESS infrastructure are already deployed. A more holistic techno-economic analysis that incorporates capital investment and maintenance costs remains an important direction for future research.

4. Conclusions

In this paper, we proposed a distributionally robust chance-constrained optimal dispatch strategy for MVDC-linked distribution networks with BESS integration. The strategy effectively handles uncertainty in PV and load forecasts through a tractable DRCCO framework.

The simulation results demonstrated that the proposed method achieves up to 44.7% cost improvement and maintains voltage reliability above 96.8% under various uncertainty scenarios. The impact of key DRCC parameters—such as the confidence level and ambiguity set radius—was analyzed, revealing a clear trade-off between operational efficiency and robustness.

However, some limitations remain. The computational time increases significantly with larger sample sizes, and the solution quality depends on the accuracy of historical forecast data. Additionally, the current model assumes that forecast errors have a zero mean, which may not always hold in real-world scenarios. If forecast bias exists, it could affect both the cost estimation and system reliability. Future research may focus on improving the computational efficiency, extending the method to multi-time-step or real-time applications and incorporating non-zero mean error models or bias-correction techniques to address potential forecast bias. Furthermore, more detailed modeling of converter dynamics could further enhance the practical applicability.

The proposed DRCCO framework can be also extended to accommodate other controllable assets such as smart inverters, demand response resources, and EV charging stations. This would involve adding new decision variables and constraints specific to each asset type, updating sensitivity coefficients, and incorporating additional uncertainty models where necessary.

Author Contributions: Conceptualization, C.H.; methodology, C.H.; software, C.H.; validation, C.H., S.S. and J.L.; writing—original draft preparation, C.H.; writing—review and editing, S.S. and J.L. All authors have read and agreed to the published version of the manuscript.

Funding: This work was supported by the research grant of the Gyeongsang National University in 2023 and by the National Research Foundation of Korea (NRF) grant funded by the Korean government (MSIT) (No. RS-2024-00451078).

Data Availability Statement: Data are contained within the article.

Conflicts of Interest: The authors declare no conflicts of interest.

Nomenclature

Sets	
k	Index of DNs
i, j	Index of buses in each DN
t	Time index
Parameters	
$(1-\alpha_V)$	Bus voltage confidence level
ϵ	Radius of ambiguity set
$P^k_{load.j.t}$, $Q^k_{load.j.t}$	Active/Reactive power of load demand at bus j, time t, in k-th DN
$P^k_{pv.t}$, $Q^k_{pv.t}$	Active/Reactive power of PV generation at bus j, time t, in k-th DN
V_{min}, V_{max}	Lower/Upper bus voltage limits at buses
SOC_{min}, SOC_{max}	Minimum/Maximum state-of-charge of BESS
r^k_{ij}, x^k_{ij}	Resistance/Reactance of branch (i, j) at k-th DN
S_{conv}	Rated apparent power capacity of MVDC link converter
S_{bess}	Rated apparent power capacity of BESS converter
$C_{conv.loss}$, $C_{bess.loss}$	Loss coefficients of MVDC converter and BESS, respectively
$C_{s.t}$	Unit electricity cost at time t

Decision Variables	
$P_{conv.t}^{k}$, $Q_{conv.t}^{k}$	Active/Reactive power outputs of MVDC link converter at time t, k-th DN
$P_{bess.t}^{k}$, $Q_{bess.t}^{k}$	Active/Reactive power outputs of BESS at time t, k-th DN
$SOC_{(t)}^{k}$	State-of-charge of BESS at time t, k-th DN
$V_{i.t}^{k}$	Voltage magnitude at bus i, time t, k-th DN
$P_{ij.t}^{k}$, $Q_{ij.t}^{k}$	Active/Reactive power flow on branch (i, j), time t, k-th DN
$l_{i.t}^{k}$	Squared current magnitude on branch (i, j), time t, k-th DN
Uncertainty Variables	
$\xi_{load.t}^{k}$	Load demand forecast error at time t, k-th DN
$\xi_{pv.t}^{k}$	PV output forecast error at time t, k-th DN
Other Variables	
$\tilde{P}_{load.j.t}^{k}$, $\tilde{P}_{pv.j.t}^{k}$	Actual load demand and PV generation after applying uncertainty
$\tilde{V}_{i.t}^{k}$	Actual bus voltage considering uncertainties and control actions
$\rho_{n.i.t}, \delta_{t.k}, \theta_t, \lambda_t$	Auxiliary variables for DRCCO formulation

References

1. Kok, C.L.; Fu, X.; Koh, Y.Y.; Teo, T.H. A Novel Portable Solar Powered Wireless Charging Device. *Electronics* **2024**, *13*, 403. [CrossRef]
2. Konstantelos, I.; Giannelos, S.; Strbac, G. Strategic Valuation of Smart Grid Technology Options in Distribution Networks. *IEEE Trans. Power Syst.* **2017**, *32*, 1293–1303.
3. Rahmani, S.; Amjady, N.; Shah, R. Application of Deep Learning Algorithms for Scenario Analysis of Renewable Energy-Integrated Power Systems: A Critical Review. *Electronics* **2025**, *14*, 2150. [CrossRef]
4. Cavus, M. Advancing Power Systems with Renewable Energy and Intelligent Technologies: A Comprehensive Review on Grid Transformation and Integration. *Electronics* **2025**, *14*, 1159. [CrossRef]
5. Quizhpe, K.; Arévalo, P.; Ochoa-Correa, D.; Villa-Ávila, E. Optimizing Microgrid Planning for Renewable Integration in Power Systems: A Comprehensive Review. *Electronics* **2024**, *13*, 3620. [CrossRef]
6. Friman, H.; Banner, I.; Sitbon, Y.; Einav, Y.; Shaked, N. Preparing the Public Opinion in the Community to Accept Distributed Energy Systems and Renewable Energy. *Energies* **2022**, *15*, 4226. [CrossRef]
7. Zhang, X.; Son, Y.; Choi, S. Optimal Scheduling of Battery Energy Storage Systems and Demand Response for Distribution Systems with High Penetration of Renewable Energy Sources. *Energies* **2022**, *15*, 2212. [CrossRef]
8. Walling, R.A.; Saint, R.; Dugan, R.C.; Burke, J.; Kojovic, L.A. Summary of distributed resources impact on power delivery systems. *IEEE Trans. Power Deliv.* **2008**, *23*, 1636–1644. [CrossRef]
9. Palmintier, B.; Broderick, R.; Mather, B.; Coddington, M.; Baker, K.; Ding, F.; Reno, M.; Lave, M.; Bharatkumar, A. On the Path to Sunshot: Emerging Issues and Challenges in Integrating Solar with the Distribution System. National Renewable Energy Laboratory; 2016. Available online: http://www.nrel.gov/docs/fy16osti/65331.pdf (accessed on 1 May 2016).
10. Salimon, S.A.; Adepoju, G.A.; Adebayo, I.G.; Howlader, H.O.R.; Ayanlade, S.O.; Adewuyi, O.B. Impact of distributed generators penetration level on the power loss and voltage profile of radial distribution networks. *Energies* **2023**, *16*, 1943. [CrossRef]
11. Zsiborács, H.; Baranyai, N.H.; Vincze, A.; Zentkó, L.; Birkner, Z.; Máté, K.; Pintér, G. Intermittent renewable energy sources: The role of energy storage in the European power system of 2040. *Electronics* **2019**, *8*, 729. [CrossRef]
12. Niu, X.; Qiu, R.; Liu, S.; Chow, X. DC-Link Voltage Fluctuation Suppression Method for Modular Multilevel Converter Based on Common-Mode Voltage and Circulating Current Coupling Injection under Unbalanced Grid Voltage. *Electronics* **2024**, *13*, 3379. [CrossRef]
13. Montoya, O.D.; Gil-González, W.; Arias-Londoño, A.; Rajagopalan, A.; Hernández, J.C. Voltage Stability Analysis in Medium-Voltage Distribution Networks Using a Second-Order Cone Approximation. *Energies* **2020**, *13*, 5717. [CrossRef]
14. Kang, K.-P.; Cho, Y.; Kim, H.-S.; Baek, J.-w. DC-Link Capacitor Voltage Imbalance Compensation Method Based Injecting Harmonic Voltage for Cascaded Multi-Module Neutral Point Clamped Inverter. *Electronics* **2019**, *8*, 155. [CrossRef]
15. Simiyu, P.; Xin, A.; Wang, K.; Adwek, G.; Salman, S. Multiterminal Medium Voltage DC Distribution Network Hierarchical Control. *Electronics* **2020**, *9*, 506. [CrossRef]
16. Chen, Y.; Yu, S.; Wang, Y.; Yang, R.; Cheng, X. An Improved AC-Link Voltage Matching Control for the Multiport Modular Multilevel DC Transformer in MVDC Applications. *Energies* **2024**, *17*, 1346. [CrossRef]

17. Blázquez, A.; Pérez-Molina, M.J.; Larruskain, D.M.; Iturregi, A.; Eguia, P. Fault Detection Algorithms in Medium-Voltage Direct-Current (MVDC) Grids. *Appl. Sci.* **2024**, *14*, 11052. [CrossRef]
18. Cheng, Q.; Li, X.; Zhang, M.; Fei, F.; Shi, G. Optimal configuration strategy of soft open point in flexible distribution network considering reactive power sources. *Energies* **2025**, *18*, 529. [CrossRef]
19. Zhao, J.; Yao, M.; Yu, H.; Song, G.; Ji, H.; Li, P. Decentralized voltage control strategy of soft open points in active distribution networks based on sensitivity analysis. *Electronics* **2020**, *9*, 295. [CrossRef]
20. Alanazi, M. Optimal integration of distributed generators and soft open points in radial distribution networks: A hybrid WCA-PSO approach. *Processes* **2025**, *13*, 1775. [CrossRef]
21. Li, J.; Zhang, Y.; Lv, C.; Liu, G.; Ruan, Z.; Zhang, F. Coordinated planning of soft open points and energy storage systems to enhance flexibility of distribution networks. *Appl. Sci.* **2024**, *14*, 8309. [CrossRef]
22. Diaaeldin, I.M.; Abdel Aleem, S.H.E.; El-Rafei, A.; Abdelaziz, A.Y.; Zobaa, A.F. Enhancement of hosting capacity with soft open points and distribution system reconfiguration: Multi-objective bilevel stochastic optimization. *Energies* **2020**, *13*, 5446. [CrossRef]
23. Ji, H.; Wang, C.; Li, P.; Ding, F.; Wu, J. Robust operation of soft open points in active distribution networks with high penetration of photovoltaic integration. *IEEE Trans. Sustain. Energy* **2019**, *10*, 280–289. [CrossRef]
24. Sun, F.; Ma, J.; Yu, M.; Wei, W. Optimized two-time scale robust dispatching method for the multi-terminal soft open point in unbalanced active distribution networks. *IEEE Trans. Sustain. Energy* **2021**, *12*, 587–598. [CrossRef]
25. Chen, T.; Song, Y.; Hill, D.J.; Lam, A.Y. Chance-constrained OPF in droop-controlled microgrids with power flow routers. *IEEE Trans. Smart Grid* **2022**, *13*, 2601–2613. [CrossRef]
26. Roald, L.; Andersson, G. Chance-constrained AC optimal power flow: Reformulations and efficient algorithms. *IEEE Trans. Power Syst.* **2018**, *33*, 2906–2918. [CrossRef]
27. Narimani, M.R.; Azizivahed, A.; Naderi, E. An Efficient Scenario-Based Stochastic Energy Management of Distribution Networks with Distributed Generation, PV Module, and Energy Storage. *arXiv* **2019**, arXiv:1910.07109.
28. Li, P.; Ji, H.; Wang, C.; Zhao, J.; Song, G.; Ding, F.; Wu, J. Coordinated Control Method of Voltage and Reactive Power for Active Distribution Networks Based on Soft Open Point. *IEEE Trans. Sustain. Energy* **2017**, *8*, 1430–1442. [CrossRef]
29. Azizivahed, A.; Gholami, K.; Arefi, A.; Arif, M.T.; Haque, M.E. Utilizing Soft Open Points for Effective Voltage Management in Multi-Microgrid Distribution Systems. *Electricity* **2024**, *5*, 1008–1021. [CrossRef]
30. Attanasio, F.; Wasterlain, S.; Pidancier, T.; Marchesoni, M.; Favre-Perrod, P.; Carpita, M. Low Voltage Soft Open Point with Energy Storage: System Simulation and Prototype Preliminary Test Results. In Proceedings of the 2018 International Symposium on Power Electronics, Electrical Drives, Automation and Motion (SPEEDAM), Amalfi, Italy, 20–22 June 2018; pp. 254–261.
31. Sarantakos, I.; Peker, M.; Zografou-Barredo, N.-M.; Deakin, M.; Patsios, C.; Sayfutdinov, T.; Taylor, P.C.; Greenwood, D. A Robust Mixed-Integer Convex Model for Optimal Scheduling of Integrated Energy Storage–Soft Open Point Devices. *IEEE Trans. Smart Grid* **2022**, *13*, 4072–4087. [CrossRef]
32. Risi, B.-G.; Riganti-Fulginei, F.; Laudani, A. Modern techniques for the optimal power flow problem: State of the art. *Energies* **2022**, *15*, 6387. [CrossRef]
33. Zhou, Q.; Bialek, J. Simplified calculation of voltage and loss sensitivity factors in distribution networks. In Proceedings of the 16th Power Systems Computation Conference, Glasgow, UK, 14–18 July 2008; pp. 1–6.
34. Mohajerin Esfahani, P.; Kuhn, D. Data-driven distributionally robust optimization using the Wasserstein metric: Performance guarantees and tractable reformulations. *Math. Program.* **2017**, *171*, 115–166. [CrossRef]
35. Gao, R.; Kleywegt, A.J. Distributionally robust stochastic optimization with Wasserstein distance. *arXiv* **2016**, arXiv:1604.02199. [CrossRef]
36. Zhao, C.; Guan, Y. Data-driven risk-averse stochastic optimization with Wasserstein metric. *Oper. Res. Lett.* **2018**, *46*, 262–267. [CrossRef]
37. Jiang, R.; Guan, Y. Data-driven chance constrained stochastic program. *Math. Program.* **2016**, *158*, 291–327. [CrossRef]
38. Shen, C.; Liu, H.; Wang, J.; Yang, Z.; Hai, C. Kullback–Leibler Divergence-Based Distributionally Robust Chance-Constrained Programming for PV Hosting Capacity Assessment in Distribution Networks. *Sustainability* **2025**, *17*, 2022. [CrossRef]
39. Han, C.; Cho, S.; Rao, R.R. Distributionally robust optimization-based stochastic operation strategy of soft open points in distribution networks. In Proceedings of the 2024 IEEE Conference on Technologies for Sustainability (SusTech), Portland, OR, USA, 14–17 April 2024; pp. 70–77.
40. Baran, M.E.; Wu, F.F. Network Reconfiguration in Distribution Systems for Loss Reduction and Load Balancing. *IEEE Trans. Power Deliv.* **1989**, *4*, 1401–1407. [CrossRef]

41. Baran, M.E.; Wu, F.F. Optimal Capacitor Placement on Radial Distribution Systems. *IEEE Trans. Power Deliv.* **1989**, *4*, 725–734. [CrossRef]
42. Das, D.; Kothari, D.; Kalam, A. Simple and Efficient Method for Load Flow Solution of Radial Distribution Networks. *Int. J. Electr. Power Energy Syst.* **1995**, *17*, 335–346. [CrossRef]

electronics

MDPI

Article

A Passivity-Based Control Integrated with Virtual DC Motor Strategy for Boost Converters Feeding Constant Power Loads

Mingyang Ou [1], Pingping Gong [1,*], Huajie Guo [2] and Gaoxiang Li [1]

1 College of Electrical Engineering, Guangxi University, Nanning 530004, China; 2312392098@st.gxu.edu.cn (M.O.); ligaoxiang@gxu.edu.cn (G.L.)
2 School of Electrical Engineering, Guangxi Vocational College of Water Resources and Electric Power, Nanning 530023, China; 13688590932@163.com
* Correspondence: gongpp@gxu.edu.cn

Abstract

This article proposes a nonlinear control strategy to address the voltage instability issue caused by the boost converter with an uncertain constant power load (CPL). This strategy combines a passivity-based controller (PBC) with a virtual DC motor controller (VDCM). Initially, a PBC is designed for the boost converter, which enhances the robustness of the converter with CPL perturbations in the DC bus voltage. To overcome the limitations of PBC, including steady-state errors resulting from variations in load or input voltage, the VDCM is incorporated, simulating the characteristics of a DC motor. This addition improves the system's inertia and damping, making it more stable and significantly enhancing its dynamic performance. The efficacy and stability analysis of the proposed control strategy is validated through both simulation and experimentation.

Keywords: passivity-based control (PBC); virtual DC motor control (VDCM); constant power load (CPL); boost converter

1. Introduction

The continuous deterioration of the natural environment, the increasing scarcity of traditional fossil fuels, and the advancement of power electronics have led to widespread attention and research on DC/DC conversion as a major form of power conversion [1]. Boost converters, in particular, are widely employed in applications that require stepping up low voltages to higher levels, such as renewable energy systems, electric vehicles, and portable electronic devices [2]. However, constant power loads (CPLs), prevalent in DC systems, exhibit complex impedance characteristics that can negatively impact power quality, leading to voltage oscillations and reduced damping [3]. This phenomenon is a major contributor to system instability.

To address this issue, many linear control strategies have been proposed and can be broadly divided into two categories: passive damping methods and active damping methods [4]. Passive damping methods primarily eliminate the impact of CPLs by incorporating hardware devices such as capacitors, resistors, and inductors [5], but this method increases the cost and size of the system, making it less efficient. Active damping methods stabilize the system by adding virtual resistance [6,7], virtual impedance control [8], and so on. However, these linear methods are limited by their reliance on small-signal models near the operating point and may fail when large perturbations occur, leading to instability. Therefore, advanced nonlinear control techniques need to be introduced. Popular nonlinear approaches include model predictive control (MPC), backstepping control, sliding mode control (SMC), and passivity-based control (PBC). A pseudo-extended Kalman filter was introduced in [9] to stabilize microgrid with CPL using stochastic nonlinear MPC, but the computation is complex and dependent on model accuracy. A backstepping controller based on droop control was used in [10] to mitigate the CPL problem. However, the design process is cumbersome and relies on a nonlinear disturbance observer, which is sensitive to system parameters. Additionally, two fast sliding mode-based controllers were proposed in [11] for buck converters with CPLs, based on an output voltage regulation approach and output power regulation scheme, respectively. Nevertheless, these require a high switching frequency of the converter. In contrast to other nonlinear control strategy methods, PBC stands out for its simplicity, highly efficient, easy realization, and thus has become one of the most effective applied nonlinear techniques [12]. PBC achieves a global asymptotically stable equilibrium by injecting a virtual resistance matrix in a specific way and reshaping the dissipated energy of the system, which is a crucial operation for system control. Thereby, it ensures the stability in the closed-loop control system and has the advantage of enabling the flexible plug-and-play operation of distributed power supplies [13–21]. A composite current-constrained controller was proposed in [13], which incorporates perturbation estimation and a nonlinear penalty term into the passivity-based control law. In [15], an alternating component passivity-based controller was designed by damping and interconnection injection.

While PBC is effective, the limitation is that it cannot eliminate the output voltage steady state errors caused by load or power supply variations [16]. To solve this problem, a nonlinear disturbance observer (NDO) has been proposed to compensate for the steady-state error generated by PBC [17,18,22], but this requires high computational resources and real-time performance. Additionally, the adaptive extended Kalman filter (AEKF) has been designed in [19], which relies heavily on priori information. Both approaches utilize load information to address this issue, thereby increasing algorithmic complexity. In comparison, control methods that eliminate steady-state error without requiring load information typically combine PBC with other control algorithms. For example, a complementary proportional-integral (PI) controller has been proposed in conjunction with an adaptive interconnection and damping assignment passivity-based controller (IDA-PBC) [20], but the control process is lengthy and prone to overshoot. Literature [21] combines integral sliding mode control (ISMC) with PBC. However, it fails to address the inherent drawbacks of sliding mode control. Furthermore, the high-speed response of the front-end power electronic converter can threaten the stable operation of the DC system when disturbances occur due to the frequent switching of distributed power sources or loads. Inspired by virtual synchro-nous motor control (VSG) in AC microgrids [23], virtual DC motor (VDCM) control has been proposed to provide inertia and damping [24]. Ref. [25] compares the control model of a DC motor with the conventional double closed-loop control to highlight the performance of the VDCM. The VDCM contains the DC motor armature equations and the mechanical rotation equations. It is known through the mechanical rotation equations that the control contains an integral part, which usually serves to eliminate the steady state error of the system.

In summary, the traditional PBC can effectively suppress the oscillation caused by the CPL, but the issue of steady-state error in its control performance remains insufficiently studied. To address this issue, this paper proposes a robust control strategy for boost converter against CPL perturbations, which combines PBC and VDCM. The Euler–Lagrange (EL) model of the boost converter is employed to design PBC, ensuring system stability and passivity. This approach enables flexible plugging and unplugging of distributed power supplies in microgrids, with better adaptability to actual circuit topologies and a more implementable structure. The VDCM serves as a patch for PBC, compensating for steady-state errors caused by various perturbations and increasing the system's inertia and damping to improve its robustness. The efficacy of the proposed method has been validated through simulations using MATLAB/Simulink R2022a software and experiments on the dSPACE1104 experimental platform. The results demonstrate that the proposed method not only possesses robustness similar to that of PBC but also eliminates steady-state errors through VDCM, addressing the limitations of PBC.

The remainder of this paper is structured in the following manner: Section 2 lays out the system model, and Section 3 delves into the details of the closed-loop control design for the PBC with VDCM. The simulation and experimental sessions are then described in Section 4. Lastly, Section 5 presents a summary of the work.

2. System Configuration and Modeling

The topology employed in this paper, where a CPL is loaded through a boost converter, is shown in Figure 1. Typical examples of CPL include a DC/DC converter connected to resistors and a DC/AC inverter driving a motor [26], where the system may oscillate and become unstable due to frequent switching of the switching devices. Assuming that the converter operates in a continuous conduction mode (CCM), the average state equation can be derived as follows:

$$\begin{cases} L\frac{di_L}{dt} = u_{in} - (1-\mu)u_C \\ C\frac{du_C}{dt} = (1-\mu)i_L - \frac{P_{CPL}}{u_C} \end{cases}, \tag{1}$$

where u_{in} represents the input voltage, L is the inductance, C is the high-voltage side capacitor, i_L denotes the inductance current of the converter, u_C is the capacitor voltage on the high-voltage side of the converter, P_{CPL} is the power of CPL, and μ is the duty cycle.

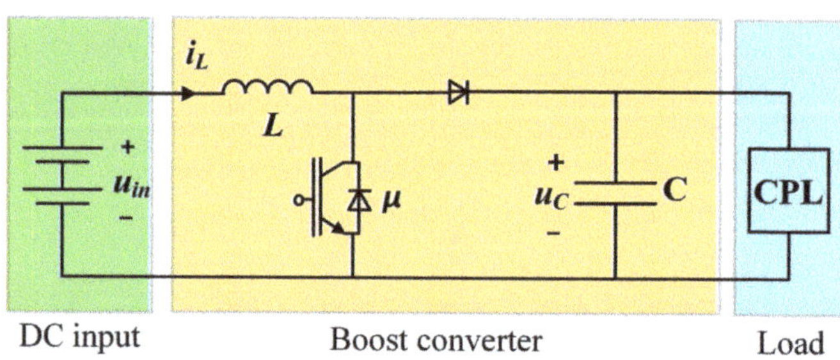

Figure 1. Boost converter with CPL model.

Define

$$X = \begin{bmatrix} x_1 \\ x_2 \end{bmatrix} = \begin{bmatrix} i_L \\ u_C \end{bmatrix}, A = \begin{bmatrix} L & 0 \\ 0 & C \end{bmatrix}, B = \begin{bmatrix} 0 & (1-\mu) \\ -(1-\mu) & 0 \end{bmatrix},$$
$$R = \begin{bmatrix} 0 & 0 \\ 0 & \frac{P_{CPL}}{u_C^2} \end{bmatrix}, U = \begin{bmatrix} u_{in} \\ 0 \end{bmatrix}.$$

Equation (1) can be rewritten in the EL equation as follows, thus establishing the basis for the controller design detailed in the subsequent chapter:

$$A\dot{X} + (B+R)X = U. \tag{2}$$

The proposed controller is designed to regulate the output voltage of the boost converter feeding CPL and better address the impacts of CPL and input voltage disturbances, thus achieving the following objectives:

$$\lim_{t\to\infty}[X - X_d] = 0, \tag{3}$$

where $X_d = \begin{bmatrix} x_{1d} \\ x_{2d} \end{bmatrix} = \begin{bmatrix} I_L \\ U_C \end{bmatrix}$ is the reference point value.

3. Design of Virtual DC Motor Compensated Passivity-Based Control

To address the instability caused by CPLs, a passive-based controller has been designed, and its stability has been analyzed. Furthermore, VDCM has been incorporated to increase the system's inertia and eliminate steady-state errors. The overall framework of the proposed method is shown in Figure 2.

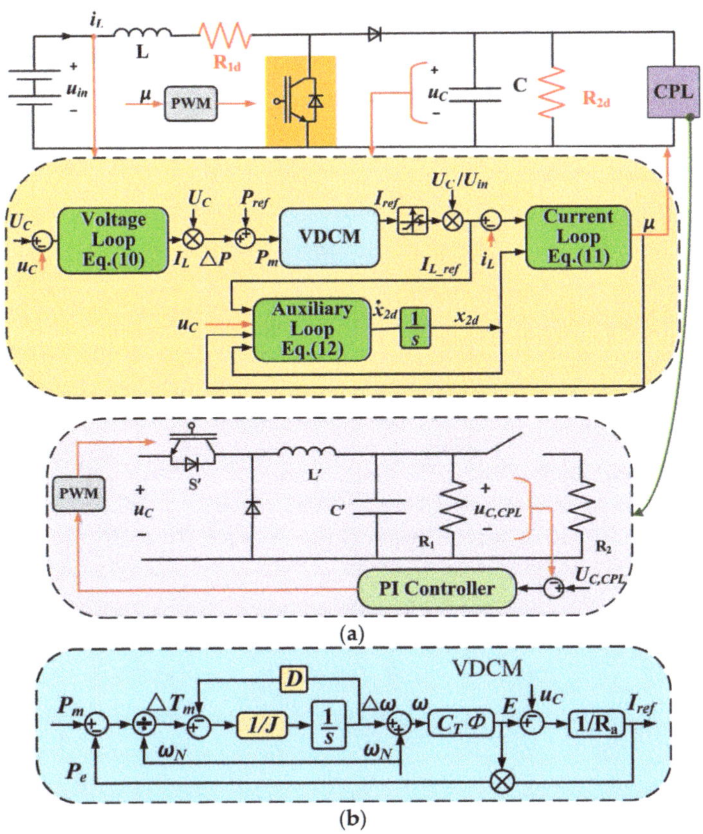

Figure 2. Block diagram of proposed control strategy: (**a**) control block diagram; (**b**) virtual DC machine control part.

3.1. Design of Passivity-Based Control

To mitigate the extreme cycle oscillation caused by CPL, the PBC applied to the boost converter was implemented by injecting virtual damping resistance, as shown in Figure 2a. This approach adds two virtual resistances to the placement of the converter. In this sense, the system is entirely passive and stable. Additionally, the CPL was configured as a Buck converter under PI control with resistive loads. The duty cycle output by the controller is compared with a triangular wave to generate a PWM signal, which controls the switching devices of the converter. The design of passivity-based controllers undergoes two necessary stages [17,18]:

I. Energy shaping stage:

Let $X = X_d + \widetilde{X}$, where $\widetilde{X}$ is the systematic error deviating from the reference point when the system oscillates. According to Equation (2), the equation can be obtained as follows:

$$A\left(\dot{X}_d + \dot{\widetilde{X}}\right) + (B+R)\left(X_d + \widetilde{X}\right) = U. \tag{4}$$

Equation (4) is expressed as follows:

$$A\dot{\widetilde{X}} + (B+R)\widetilde{X} = U - \left[A\dot{X}_d + (B+R)X_d\right]. \tag{5}$$

II. Damping injection stage:

By injecting the virtual damping matrix R_d into Equation (5), it can be concluded that

$$A\dot{\widetilde{X}} + (B+R_Z)\widetilde{X} = U - \left[A\dot{X}_d + (B+R)X_d - R_d\widetilde{X}\right], \tag{6}$$

where

$$R_Z = \begin{bmatrix} R_{1d} & 0 \\ 0 & \frac{1}{R_{2d}} + \frac{P_{CPL}}{u_C^2} \end{bmatrix},$$
$$R_d = R_Z - R = \begin{bmatrix} R_{1d} & 0 \\ 0 & \frac{1}{R_{2d}} \end{bmatrix}.$$

By adding virtual resistors, the transient energy dissipation and Lyapunov stability can be ensured. The series resistance of the inductor circuit (R_{1d}) is sufficient to ensure energy dissipation in the inductor and effectively suppress inductor current ripple. The virtual resistance in the parallel capacitor circuit (R_{2d}) is sufficiently small to minimize energy dissipation in the capacitor and suppress output voltage ripple. It should be noted that the case of $u_C = 0$ generally occurs only during equipment startup, when the load side has no voltage and the CPL does not activate, thus exerting no impact on the controller. In practical operation, protective actions will be triggered to shut down the converter once the load voltage drops below a specific threshold [27]. Therefore, $\widetilde{X} = 0$, and Equation (6) can be rewritten as

$$U - \left[A\dot{X}_d + (B+R)X_d - R_d\widetilde{X}\right] = 0. \tag{7}$$

Rearrange Equation (7) and get the following equations:

$$u_{in} - L\dot{x}_{1d} - (1-\mu)x_{2d} + R_{1d}(x_1 - x_{1d}) = 0, \tag{8}$$

$$-C\dot{x}_{2d} + (1-\mu)x_{1d} - \frac{P_{CPL}}{x_{2d}} + \frac{1}{R_{2d}}(x_2 - x_{2d}) = 0. \tag{9}$$

The PBC formula in Equation (8), enables accurate tracking of voltage reference value during steady-state operation. However, due to the randomness and fluctuation of CPL, steady-state errors are introduced, which further expand the output voltage error. To mitigate this issue, the VDCM approach, discussed in the subsequent section, is employed. Therefore, I_L needs to be modified as I_{L_ref} through VDCM. It should be noted that I_L represents the reference voltage generated by the PBC voltage control loop, while I_{L_ref} corresponds to the adjusted current reference value obtained by processing I_L through the VDCM compensation module. In addition, the boost converter of direct control via inner and outer loops is ineffective in regulating the DC bus voltage (x_2) to a stable equilibrium point due to the non-minimum phase characteristics [18]. Consequently, auxiliary loop control is necessary [17,28]. The duty cycle can be expressed as $1-\mu = u_{in}/U_C$, and according to Equations (8) and (9), the duty cycle μ and the reference value of inductance current of PBC I_L are defined as follows

$$I_L = \frac{P_{CPL}}{u_{in}} + \frac{U_C}{R_{2d}u_{in}}(U_C - u_C), \tag{10}$$

$$\mu = 1 - \frac{1}{x_{2d}}\left[u_{in} - R_{1d}\left(I_{L\_ref} - i_L\right)\right]. \tag{11}$$

According to Equation (9), the auxiliary controller can be obtained as

$$\dot{x}_{2d} = \frac{1}{C}\left[-\frac{P_{CPL}}{x_{2d}} + \frac{1}{R_{2d}}(u_C - x_{2d}) + (1-\mu)I_{L\_ref}\right]. \tag{12}$$

3.2. Stability Analysis of PBC

Lyapunov stability analysis is commonly employed to analyze the stability of PBC. Based on Equations (6) and (7) and the positive definite matrix A, the equation and a Lyapunov function $V(x)$ can be derived:

$$A\dot{\widetilde{X}} + [B + R_Z]\widetilde{X} = 0, \tag{13}$$

$$V(x) = \frac{1}{2}\widetilde{X}^T A\widetilde{X} > 0(\forall\widetilde{x} \neq 0). \tag{14}$$

The derivative of $V(x)$ related to time can be written as

$$\dot{V}(x) = \widetilde{X}^T A\dot{\widetilde{X}}. \tag{15}$$

Based on Equation (13), $\dot{\widetilde{X}}$ can be expressed as

$$\dot{\widetilde{X}} = -A^{-1}[B + R_Z]\widetilde{X}. \tag{16}$$

The combination of Equations (15) and (16) yields

$$\dot{V}(x) = -\left[\widetilde{X}^T R_Z\widetilde{X} + \widetilde{X}^T B\widetilde{X}\right]. \tag{17}$$

Since B is an antisymmetric matrix, $\widetilde{X}^T B\widetilde{X} = 0$. If the matrix R_Z is positive, then

$$\dot{V}(x) = -\widetilde{X}^T R_Z\widetilde{X} < 0. \tag{18}$$

Therefore, the system satisfies Lyapunov stability and asymptotic stability.

3.3. Virtual DC Machine Control Strategy

As mentioned above, the inherent limitation of PBC is that its propensity to generate steady-state errors in response to disturbances in the load or power supply. To address this shortcoming, VDCM control is incorporated. This control strategy features an integrator that tracks the desired current value in real time, thereby eliminating the steady-state error. Furthermore, VDCM enhances the inertia and damping of the DC system by simulating the external characteristics of the DC motor, as shown in Figure 3. The armature current I_a is served as the reference value I_{ref} on output side of the converter. The output voltage of the DC machine U_o simulates the output voltage U_C of the boost converter.

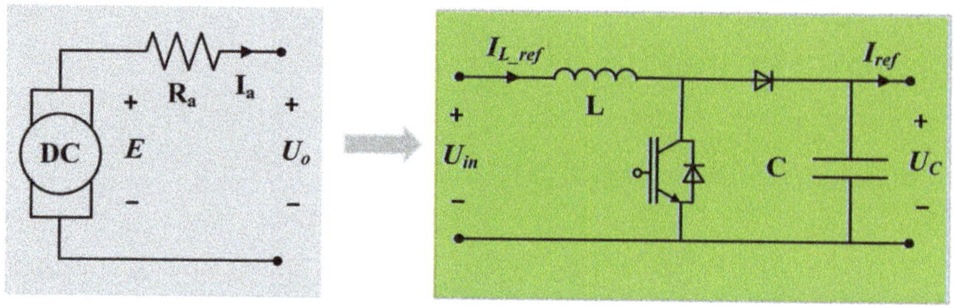

Figure 3. Concept model of VDCM strategy.

The VCDM can be mathematically modeled using the armature equation and the mechanical rotation equation of the DC motor [24], in which the mechanical rotation equation is expressed as

$$J\frac{d\omega}{dt} = T_m - T_e - D(\omega - \omega_N), \tag{19}$$

$$T_e = \frac{P_e}{\omega}, \tag{20}$$

where J represents the moment of inertia, D denotes the damping coefficient, T_m is the mechanical torque, T_e is the electromagnetic torque, ω signifies angular velocity, ω_N is the rated angular velocity, and P_e represents the electromagnetic power.

The armature equation of the DC machine is written as follows:

$$U_o = E - R_a I_a, \tag{21}$$

$$E = C_T \phi \omega, \tag{22}$$

where E is the armature induced electromotive force, R_a denotes the armature equivalent resistance, C_T is the torque coefficient, and ϕ is flux per pole.

The VDCM control block diagram is illustrated in Figure 2b, which is consistent with the aforementioned VDCM control formula. The input and output of the VDCM control are illustrated in Figure 2a. In conventional VDCM control, a voltage PI controller generates the deviation power ΔP, which is then added P_{ref} to obtain the mechanical power P_m. After passing through the VDCM block, the adjusted reference current is output, thereby enhancing the system's damping and inertia. Similarly, the reference current I_L, obtained from the voltage error via the PBC voltage loop, should undergo feedback adjustment through the VDCM. The resulting control signal I_{L_ref} is fed into the subsequent PBC current loop to eliminate steady-state voltage errors and improve system stability. After being controlled by VDCM, the reference current on output side of the converter I_{ref} is converted into the reference inductor current I_{L_ref} through the ratio U_C/U_{in}.

4. Results

4.1. Simulation Results

To validate the stability of the proposed method and its effectiveness in suppressing steady-state errors compared to traditional PBC under load variations, we apply MATLAB/Simulink R2022a software testing under CPL, reference voltage and input voltage variations. The simulation model is shown in Figure 4, employing the ode1 (Euler) solver. The reference voltage is set to 27 V. The parameters of boost converter and CPL, as well as the control parameters, are listed in Table 1. The subsequent text presents and analyzes the simulation results.

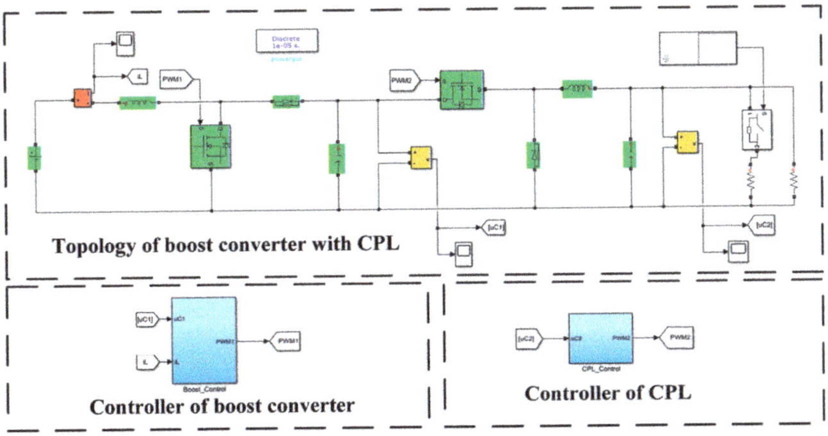

Figure 4. Boost converter with CPL in Simulink.

Table 1. Main parameters of simulation and experiment.

Boost Converter Parameters	Value
Input voltage u_{in}/V	12
Input filter inductor L/mH	2
Capacitance C/μF	1000
Switching frequency f/kHz	10
Control Parameters of Boost Converter	
PBC gains R_{1d}/Ω	10^6
PBC gains R_{2d}/Ω	0.05
Moment of inertia J/kg·m^2	0.5
Damping coefficient D	0.3
Rated angular velocity ω_N/rad·s^{-1}	2π·50
Torque coefficient C_T	18.48
Flux per pole ϕ/Wb	0.0698
Armature equivalent resistance R_a/Ω	1
Voltage loop proportional gain K_{vp}	0.6
Voltage loop integral gain K_{vi}	20
Current loop proportional gain K_{ip}	1
Current loop integral gain K_{ii}	25
CPL Parameters	
Load voltage $v_{out_{buck}}$/V	15
Load power P_{const}/W	4.5/27
Filter inductor L_f/mH	2
Output capacitance C_f/μF	100
Control Parameters of CPL	
proportional gain K_p	3
Integral gain K_i	20

4.1.1. Constant Power Load Variations Test

The simulation results of the CPL variations test are presented in Figure 5. In this test, the output voltage is regulated at 27 V, with the initial CPL of 4.5 W. At 1 s, the CPL suddenly increases to 27 W and then returns to 4.5 W at 2 s. The comparative analysis with dual-loop PI control reveals that the proposed method reaches the reference value approximately 90 ms faster, accompanied by a smaller bus voltage overshoot of around 3.1 V. Moreover, the proposed method can eliminate steady-state errors of capacitor voltage caused by load changes as shown in Figure 5a, which is a limitation of PBC. Figure 5b shows the waveform of the inductor current. It can be observed that the current ripple under the traditional PI control is larger than that under the other two control methods. Therefore, it is evident that the proposed method has stronger control stability and faster return to the reference value. To investigate the response of various parameters to the capacitor voltage, simulation results are presented with a single parameter modified while the others are held constant in Figure 6. The simulation reveals that R_{2d} in Figure 6a has a significant impact on the tracking of the reference point, and a smaller value is selected to achieve a better control effect. The effect of different values of R_{1d} in Figure 6b has lower sensitivity of the system, but closer inspection reveals that larger values result in smoother performance. Additionally, J in Figure 6c has a substantial influence on the dynamic response of the capacitor voltage, and choosing the appropriate D in Figure 6d is crucial for controlling the voltage to converge to the reference value.

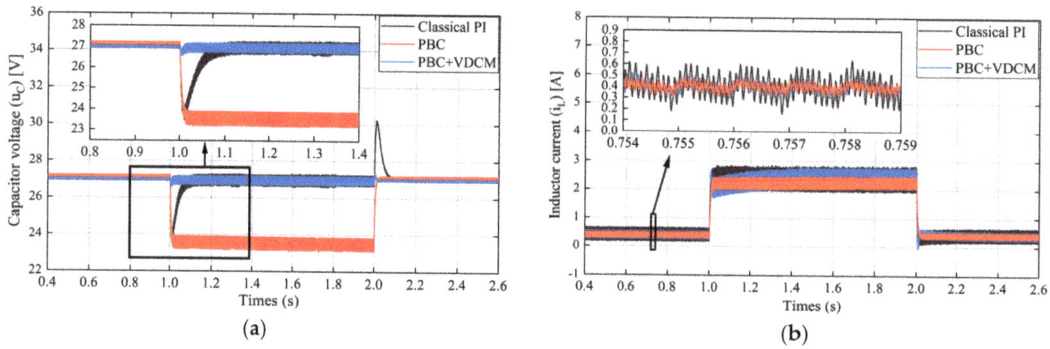

Figure 5. Comparative simulation results with sudden CPL changes: (**a**) capacitor voltage; (**b**) inductor current.

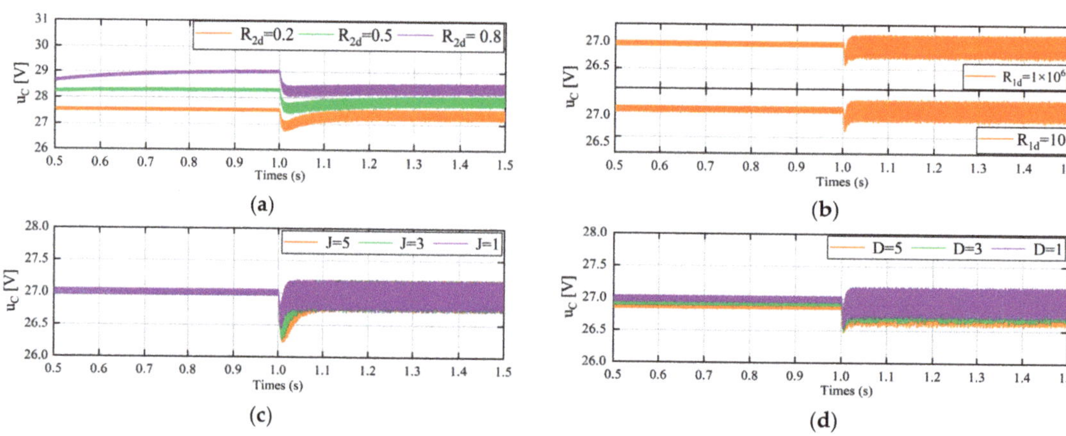

Figure 6. Dynamic response of capacitor voltage at different parameters for sudden CPL changes: (**a**) PBC gain R_{2d}; (**b**) PBC gain R_{1d}; (**c**) Moment of inertia J; (**d**) Damping coefficient D.

4.1.2. Reference Voltage Variation Test

The simulation results of the reference voltage variation test are shown in Figure 7. At 1 s, the reference voltage is stepped down from 27 V to 22 V. A comparative analysis reveals that the output voltage of both PBC and PBC + VDCM exhibits no overshoot, whereas classical PI control exhibits an overshoot of approximately 0.5 V, as shown in Figure 7a. Furthermore, the settling time of the proposed method is 53 ms less than that of PI control and 12 ms less than that of PBC. The inductor current variation amplitude of PBC + VDCM is smaller than that of the other two control methods, and its ripple is also smaller than that of classical PI control as shown in Figure 7b. In summary, the proposed method demonstrates excellent dynamic response and superior performance in terms of voltage and current regulation.

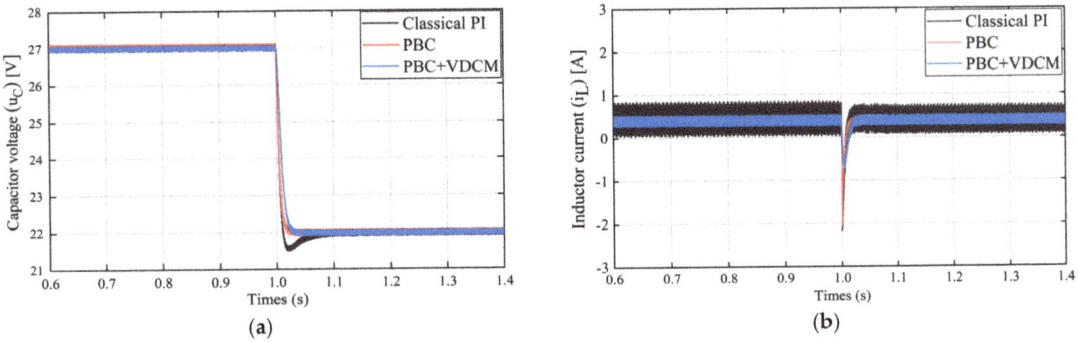

Figure 7. Simulation results with sudden reference voltage change: (**a**) Capacitor voltage; (**b**) Inductor current.

4.1.3. Input Voltage Variation Test

The simulation results of the input voltage variation experiment are shown in Figure 8. In this test, the system carries a CPL of 4.5 W, and the input voltage is ramped up from 12 V to 18 V in 1 s. The simulation results demonstrate that the PBC + VDCM strategy exhibits superior performance. Similarly to the PI controller, it eliminates steady-state error, while its overshoot is approximately 0.5 V smaller and its settling time is roughly 109 ms shorter, as shown in Figure 8a. In contrast, the PBC fails to eliminate steady-state error. Furthermore, the inductor current ripple is minimized, and the current reaches its stabilization point quickly as illustrated in Figure 8b, highlighting the excellent dynamic response of the proposed method.

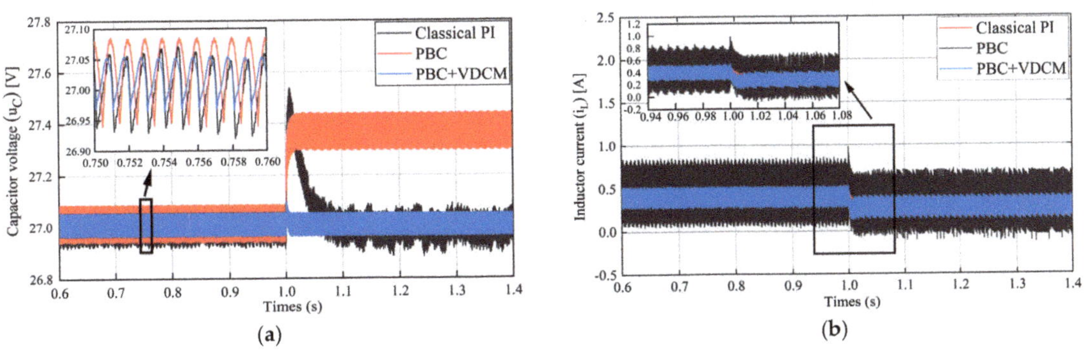

Figure 8. Simulation results with sudden input voltage change: (**a**) capacitor voltage; (**b**) inductor current.

4.1.4. Phase Plot

Given that the proposed control method is nonlinear and the boost converter is a second-order system, a global view of the proposed controller's behavior can be obtained by phase plot. Figure 9 shows the phase plot of the Boost converter with a 27 W CPL under the proposed control method. The state variables, namely the inductor current and capacitor voltage, are plotted as time-varying curves. Different colors are used to represent different initial states of the state variables. It can be seen that the phase trajectories fitted by the two state variables under different initial conditions remain within the region of attraction and converge to the equilibrium point, thereby proving the stability of the system.

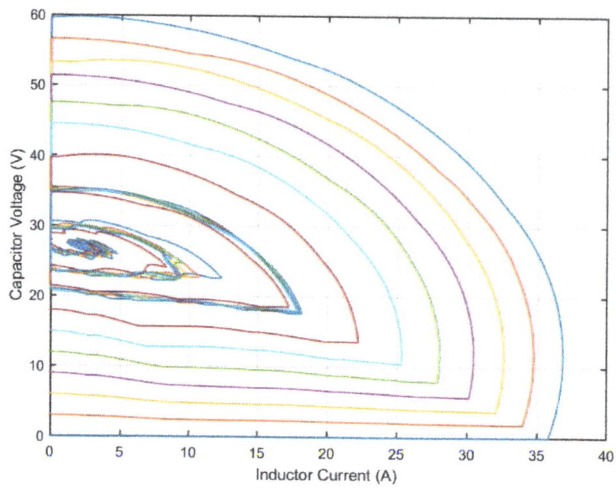

Figure 9. Phase plots of the system with the proposed method.

4.2. Experimental Results

To further validate the effectiveness of the proposed method, a boost converter with CPL experimental platform is constructed, based on the dSPACE1104, as shown in Figure 10. Multiple DC power supplies are employed to drive the circuit board, supply power to the converter's input side, and facilitate experiments on power supply disturbances. The current and voltage of the converter are sampled into dSPACE via data lines. dSPACE runs the algorithm and outputs the control variables to the PWM generator, which generates PWM signals to control the converter's switching devices. A computer is utilized to initialize dSPACE and adjust the control parameters. Voltage and current waveforms are monitored using an oscilloscope. The output voltage of the boost converter is controlled at 27 V. The converter experiment utilized the same parameter set as the simulation, as listed in Table 1.

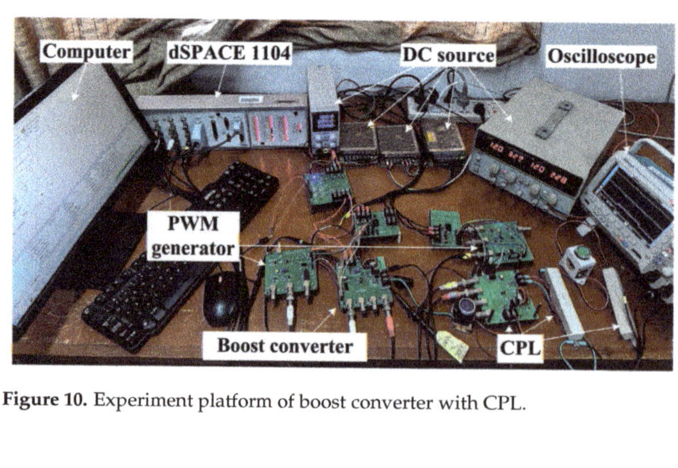

Figure 10. Experiment platform of boost converter with CPL.

4.2.1. Experimental Test of CPL Variations

Figure 11 illustrates the comparative experimental results of the boost converter under different control strategies for the constant power load, which transitions from 4.5 W to 27 W and back to 4.5 W. As depicted in Figure 11a, under classical PI control, the output voltage waveform (u_C) exhibits a noticeable drop and a relatively long transient process following a sudden load change before returning to the reference value. Figure 11b demonstrates that under PBC, the voltage drops considerably after the load increase, and the steady-state error remains uncompensated. In contrast, Figure 11c shows that the voltage fluctuation is essentially negligible under the proposed control strategy. The dynamic and steady-state performance of the boost converter is significantly enhanced by the proposed control, attributable to its PBC and VDCM characteristics.

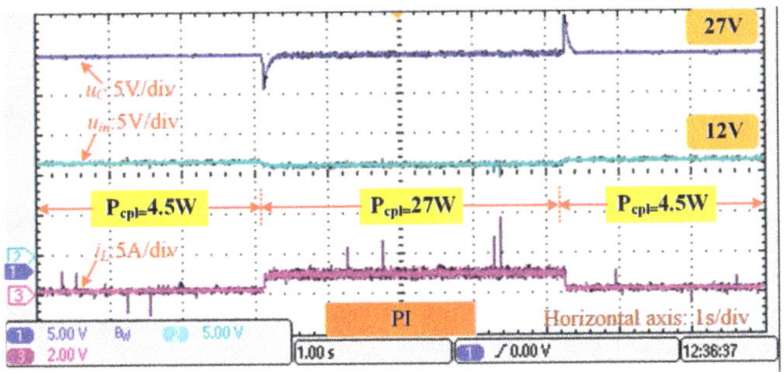

(**a**)

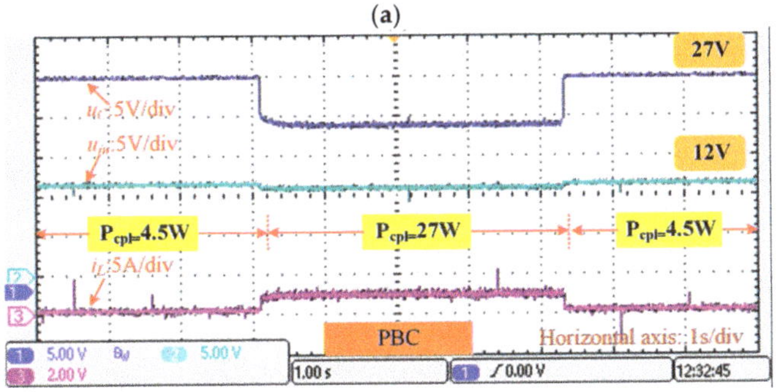

(**b**)

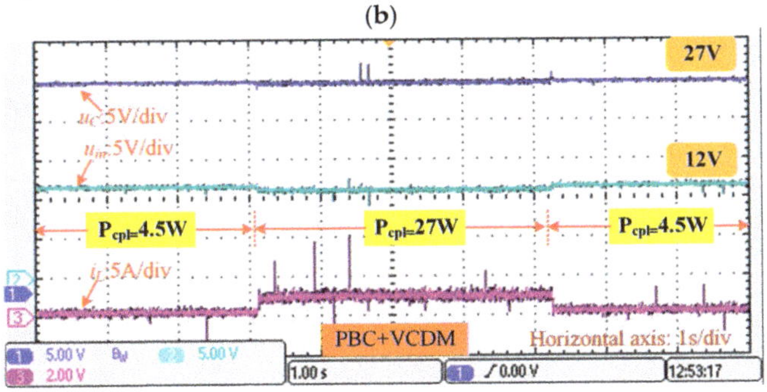

(**c**)

Figure 11. Comparative experimental results with sudden CPL changes: (**a**) PI; (**b**) PBC; (**c**) PBC + VDCM.

4.2.2. Experimental Test of Reference Voltage Variation

Figure 12 presents a comparison of experimental results illustrating the variation in reference voltage from 27 V to 22 V under different control strategies. As depicted in Figure 12a, under classical PI control, the voltage waveforms exhibit overshooting and require an extended period for stabilization. Figure 12b,c display the experimental results for the PBC and the proposed control method, respectively. The longer control time observed in the proposed control method, compared to the PBC, can be attributed to the VDCM increasing the system's damping and inertia, which enhances stability at the constant reference voltage value.

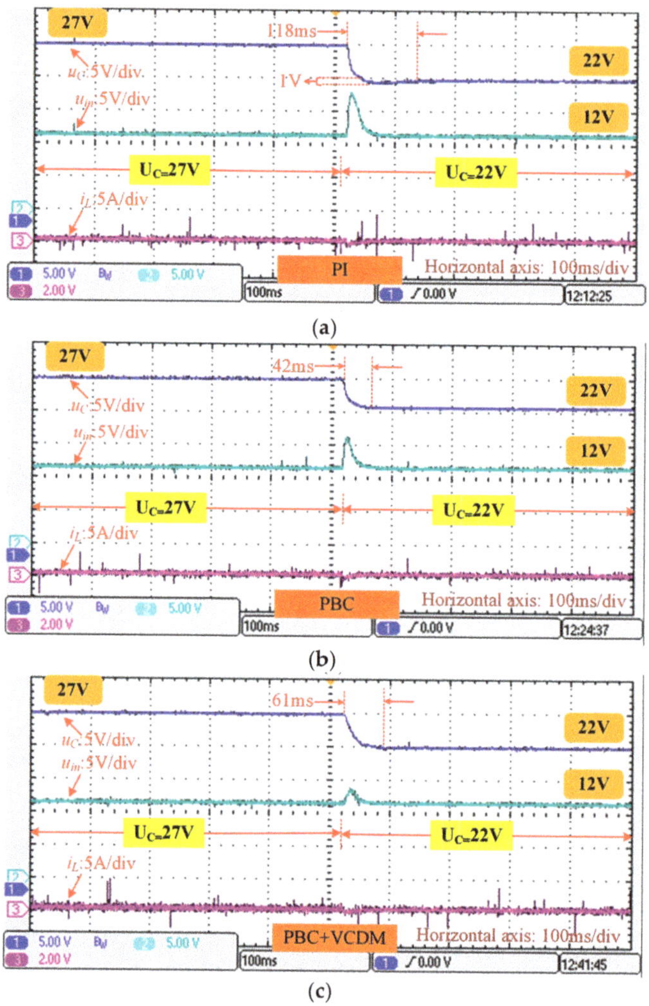

Figure 12. Comparative experimental results with sudden reference voltage change: (**a**) PI; (**b**) PBC; (c) PBC + VDCM.

4.2.3. Experimental Test of Input Voltage Variation

Figure 13 presents the experimental comparison results of the input voltage abruptly changing from 12 V to 18 V under different control strategies. The data clearly indicates that the proposed method, as depicted in Figure 13c, not only exhibits a superior response time compared to the classical PI control shown in Figure 13a, but also effectively mitigates the steady-state error associated with the PBC illustrated in Figure 13b following the input voltage change.

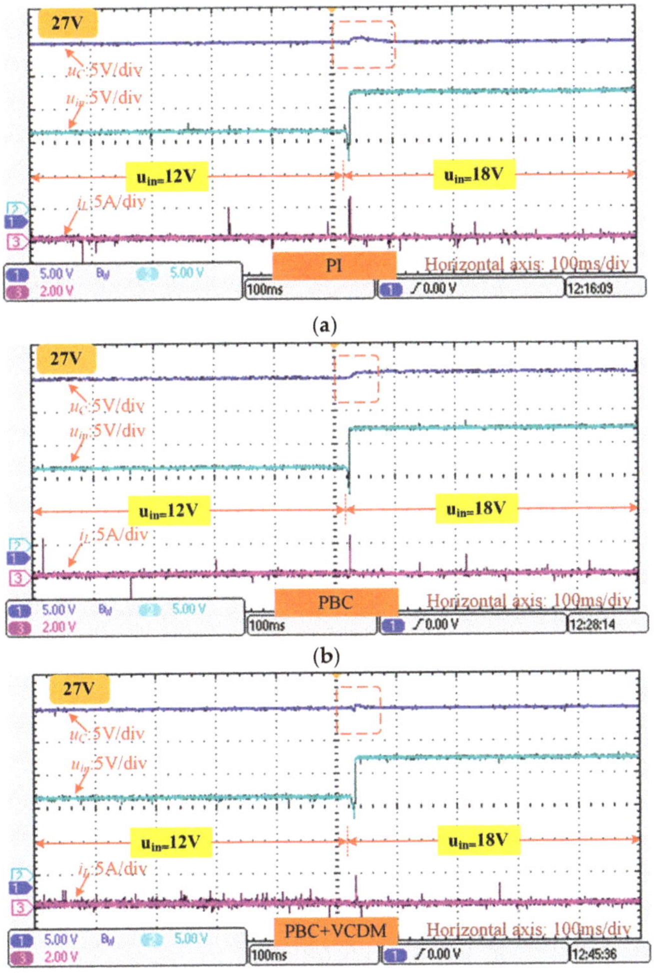

Figure 13. Comparative experimental results with sudden input voltage changes: (**a**) PI; (**b**) PBC; (**c**) PBC + VDCM.

5. Conclusions

This paper proposes a control method that combines PBC and VDCM to maintain the stability of the DC bus voltage under disturbances, such as large fluctuations in CPL. The PBC can show the good robustness and low overshoot under disturbance condition. The VDCM is applied to improve system's inertia and damping, while compensating for the steady-state error caused by PBC during load or input voltage changes. Both the simulation and experimental results demonstrate that the PBC + VDCM control method exhibits strong robustness and dynamic characteristics, characterized by small overshoots in bus voltage and inductor current, and rapid convergence to the reference value. The proposed control method has application potential in components of DC microgrids such as photovoltaics, electric vehicle charging piles, and energy storage systems.

However, the research in this paper still has certain limitations, and future research work can focus on the following two aspects:

(1) Verifying the applicability of the proposed control method in other types of DC/DC converters, such as topological structures like buck converters and bidirectional buck/boost converters;
(2) DC microgrids incorporating DC devices such as photovoltaics, energy storage systems, and electric vehicle charging piles are typical multi-converter systems. Therefore, it is necessary to further explore the adaptability and scalability of the proposed method in such multi-converter systems, encompassing aspects such as coordinated control of multiple converters and their interactive influences.

Author Contributions: Conceptualization, M.O. and P.G.; methodology, M.O.; validation, M.O.; formal analysis, P.G.; writing—original draft preparation, M.O.; supervision, P.G. and G.L.; project administration, H.G.; funding acquisition, H.G. All authors have read and agreed to the published version of the manuscript.

Funding: This research was funded by The 2024 Middle-aged and Young Teachers' Basic Ability Promotion Project of Guangxi, grant number 2024KY1143.

Data Availability Statement: The original contributions presented in this study are included in the article. Further inquiries can be directed to the corresponding author.

Conflicts of Interest: The authors declare no conflicts of interest.

Nomenclature

u_{in}, u_C	Input, capacitor (output) voltage of the converter
L	Inductance of the converter
C	Capacitor of the converter
i_L	Inductance current of the converter
P_{CPL}	Power of CPL
μ	Duty cycle of the converter
I_L, I_{L_ref}, I_{ref}	Reference inductance current of PBC, VDCM, reference current on output side of the converter
U_{in}, U_C	Reference input, capacitor (output) voltage
R_{1d}, R_{2d}	PBC gains
J	Moment of inertia in VDCM
D	Damping coefficient in VDCM
T_m, T_e	Mechanical and electromagnetic torque in VDCM
ω, ω_N	Angular velocity and rated angular velocity in VDCM
E	Armature induced electromotive force in VDCM
R_a	Armature equivalent resistance in VDCM
I_a	Armature current in VDCM
U_o	Output voltage of the DC machine in VDCM
C_T	Torque coefficient in VDCM
ϕ	Flux per pole in VDCM
P_{ref}, P_m, P_e	Reference average power of loads; mechanical power; and electromagnetic power in VDCM

References

1. Xia, Q.; Chen, J.; Yue, J.; Lyu, F.; Cui, Z. Health condition determination of DC/DC converters based on abductive learning. *Measurement* **2025**, *248*, 116910. [CrossRef]
2. Cheng, H.; Jung, S.; Kim, Y.-B. A novel reinforcement learning controller for the DC-DC boost converter. *Energy* **2025**, *321*, 135479. [CrossRef]
3. Rahimian, M.M.; Mohammadi, H.R.; Guerrero, J.M. Constant power load issue in DC/DC multi-converter systems: Past studies and recent trends. *Electr. Power Syst. Res.* **2024**, *235*, 110851. [CrossRef]
4. Singh, S.; Gautam, A.R.; Fulwani, D. Constant power loads and their effects in DC distributed power systems: A review. *Renew. Sustain. Energy Rev.* **2017**, *72*, 407–421. [CrossRef]
5. Cespedes, M.; Xing, L.; Sun, J. Constant-Power Load System Stabilization by Passive Damping. *IEEE Trans. Power Electron.* **2011**, *26*, 1832–1836. [CrossRef]
6. Wu, M.; Lu, D.D.-C. A Novel Stabilization Method of LC Input Filter with Constant Power Loads Without Load Performance Compromise in DC Microgrids. *IEEE Trans. Ind. Electron.* **2015**, *62*, 4552–4562. [CrossRef]
7. Rahimi, A.M.; Emadi, A. Active Damping in DC/DC Power Electronic Converters: A Novel Method to Overcome the Problems of Constant Power Loads. *IEEE Trans. Ind. Electron.* **2009**, *56*, 1428–1439. [CrossRef]
8. Zhang, X.; Ruan, X.; Zhong, Q.-C. Improving the Stability of Cascaded DC/DC Converter Systems via Shaping the Input Impedance of the Load Converter with a Parallel or Series Virtual Impedance. *IEEE Trans. Ind. Electron.* **2015**, *62*, 7499–7512. [CrossRef]
9. Kowsari, E.; Zarei, J.; Razavi-Far, R.; Saif, M.; Dragicevic, T.; Khooban, M.-H. A Novel Stochastic Predictive Stabilizer for DC Microgrids Feeding CPLs. *IEEE J. Emerg. Sel. Top. Power Electron.* **2021**, *9*, 1222–1232. [CrossRef]
10. Yu, D.; Zhang, W.; Chu, Z.; Zhang, H.; Wang, Z. An optimized synchronous approach to DC/DC droop backstepping control considering voltage compensation in DC microgrids. *Electr. Power Syst. Res.* **2024**, *228*, 110014. [CrossRef]
11. Martinez-Trevino, B.A.; El Aroudi, A.; Vidal-Idiarte, E.; Cid-Pastor, A.; Martinez-Salamero, L. Sliding-mode control of a boost converter under constant power loading conditions. *IET Power Electron.* **2019**, *12*, 521–529. [CrossRef]
12. Xu, Q.; Vafamand, N.; Chen, L.; Dragicevic, T.; Xie, L.; Blaabjerg, F. Review on Advanced Control Technologies for Bidirectional DC/DC Converters in DC Microgrids. *IEEE J. Emerg. Sel. Top. Power Electron.* **2021**, *9*, 1205–1221. [CrossRef]
13. Huang, S.; Guo, T.; Wang, X.; Li, S.; Li, Q. An Output Voltage Tracking Control Method with Overcurrent Protection Property for Disturbed DC-DC Boost Converters. *IEEE Trans. Circuits Syst. I-Regul. Pap.* **2024**, *71*, 1889–1900. [CrossRef]
14. He, W.; Ortega, R. Design and Implementation of Adaptive Energy Shaping Control for DC-DC Converters with Constant Power Loads. *IEEE Trans. Ind. Inform.* **2020**, *16*, 5053–5064. [CrossRef]
15. Wang, M.; Tang, F.; Wu, X.; Zhang, Y.; Wang, J. Passivation and Passivity-Based Control of DC-DC Converter with Unknown Constant Power Loads. *IEEE J. Emerg. Sel. Top. Power Electron.* **2022**, *10*, 4041–4058. [CrossRef]
16. Hassan, M.A.; Su, C.L.; Pou, J.; Sulligoi, G.; Almakhles, D.; Bosich, D.; Guerrero, J.M. DC Shipboard Microgrids with Constant Power Loads: A Review of Advanced Nonlinear Control Strategies and Stabilization Techniques. In Proceedings of the IEEE-Power-and-Energy-Society General Meeting (PESGM), Orlando, FL, USA, 16–20 July 2023.
17. Hassan, M.; Su, C.-L.; Chen, F.-Z.; Lo, K.-Y. Adaptive Passivity-Based Control of a DC-DC Boost Power Converter Supplying Constant Power and Constant Voltage Loads. *IEEE Trans. Ind. Electron.* **2022**, *69*, 6204–6214. [CrossRef]
18. Hassan, M.A.; Sadiq, M.; Hoang, L.Q.N.; Su, C.-L. Adaptive Passivity-Based Control for Bidirectional DC-DC Power Converter Supplying Constant Power Loads in DC Shipboard Microgrids. In Proceedings of the IEEE International Future Energy Electronics Conference (IFEEC), Taipei, Taiwan, 16–19 November 2021.
19. Abdolahi, M.; Adabi, J.; Mousavi, S.Y.M. An Adaptive Extended Kalman Filter with Passivity-Based Control for DC-DC Converter in DC Microgrids Supplying Constant Power Loads. *IEEE Trans. Ind. Electron.* **2024**, *71*, 4873–4882. [CrossRef]
20. Zeng, J.; Zhang, Z.; Qiao, W. An Interconnection and Damping Assignment Passivity-Based Controller for a DC-DC Boost Converter with a Constant Power Load. *IEEE Trans. Ind. Appl.* **2014**, *50*, 2314–2322. [CrossRef]
21. Badji, A.; Obeid, H.; Hilairet, M.; Laghrouche, S.; Abdeslam, D.O.; Djerdir, A. Enhanced energy management of fuel cell electric vehicles using integral sliding mode control and passivity-based control. *Appl. Energy* **2025**, *377*, 124653. [CrossRef]
22. Hassan, M.A.; Li, E.-P.; Li, X.; Li, T.; Duan, C.; Chi, S. Adaptive Passivity-Based Control of dc-dc Buck Power Converter with Constant Power Load in DC Microgrid Systems. *IEEE J. Emerg. Sel. Top. Power Electron.* **2019**, *7*, 2029–2040. [CrossRef]
23. Yang, M.; Zhang, L.; Song, X.; Kang, W.; Kang, Z. A Transient Control Strategy for Grid-Forming Photovoltaic Systems Based on Dynamic Virtual Impedance and RBF Neural Networks. *Electronics* **2025**, *14*, 785. [CrossRef]
24. Tan, S.; Dong, G.; Zhang, H.; Zhi, N.; Xiao, X. Virtual DC Machine Control Strategy of Energy Storage Converter in DC Microgrid. In Proceedings of the IEEE Electrical Power and Energy Conference (EPEC), Ottawa, ON, Canada, 12–14 October 2016.
25. Lin, G.; Ma, J.; Li, Y.; Rehtanz, C.; Liu, J.; Wang, Z.; Wang, P.; She, F. A Virtual Inertia and Damping Control to Suppress Voltage Oscillation in Islanded DC Microgrid. *IEEE Trans. Energy Convers.* **2021**, *36*, 1711–1721. [CrossRef]

26. Emadi, A.; Khaligh, A.; Rivetta, C.H.; Williamson, G.A. Constant power loads and negative impedance instability in automotive systems: Definition, modeling, stability, and control of power electronic converters and motor drives. *IEEE Trans. Veh. Technol.* **2006**, *55*, 1112–1125. [CrossRef]
27. Kwasinski, A.; Krein, P.T. Stabilization of constant power loads in dc-dc converters using passivity-based control. In Proceedings of the 29th International Telecommunications Energy Conference, Rome, Italy, 30 September–5 October 2007; pp. 867–874.
28. Sira-Ramirez, H.; Ortega, R.; Garcia-Esteban, M. Adaptive passivity-based control of average dc-to-dc power converter models. *Int. J. Adapt. Control Signal Process.* **1998**, *12*, 63–80. [CrossRef]

electronics

MDPI

Article

Symmetric Optimization Strategy Based on Triple-Phase Shift for Dual-Active Bridge Converters with Low RMS Current and Full ZVS over Ultra-Wide Voltage and Load Ranges

Longfei Cui, Yiming Zhang, Xuhong Wang * and Dong Zhang

School of Information Science and Technology, Beijing University of Technology, Beijing 100124, China; cuilongfei1@emails.bjut.edu.cn (L.C.); ymzhang@bjut.edu.cn (Y.Z.); dong_zhang@emails.bjut.edu.cn (D.Z.)

* Correspondence: wangxuhong@bjut.edu.cn

Abstract

Dual-active bridge (DAB) converters have emerged as a preferred topology in electric vehicle charging and energy storage applications, owing to their structurally symmetric configuration and intrinsic galvanic isolation capabilities. However, conventional triple-phase shift (TPS) control strategies face significant challenges in maintaining high efficiency across ultra-wide output voltage and load ranges. To exploit the inherent structural symmetry of the DAB topology, a symmetric optimization strategy based on triple-phase shift (SOS-TPS) is proposed. The method specifically targets the forward buck operating mode, where an optimization framework is established to minimize the root mean square (RMS) current of the inductor, thereby addressing both switching and conduction losses. The formulation explicitly incorporates zero-voltage switching (ZVS) constraints and operating mode conditions. By employing the Karush–Kuhn–Tucker (KKT) conditions in conjunction with the Lagrange multiplier method (LMM), the refined control trajectories corresponding to various power levels are analytically derived, enabling efficient modulation across the entire operating range. In the medium-power region, full-switch ZVS is inherently satisfied. In the low-power operation, full-switch ZVS is achieved by introducing a modulation factor λ, and a selection principle for λ is established. For high-power operation, the strategy transitions to a conventional single-phase shift (SPS) modulation. Furthermore, by exploiting the inherent symmetry of the DAB topology, the proposed method reveals the symmetric property of modulation control. The modulation strategy for the forward boost mode can be efficiently derived through a duty cycle and voltage gain mapping, eliminating the need for re-derivation. To validate the effectiveness of the proposed SOS-TPS strategy, a 2.3 kW experimental prototype was developed. The measured results demonstrate that the method ensures ZVS for all switches under the full load range, supports ultra-wide voltage conversion capability, substantially suppresses RMS current, and achieves a maximum efficiency of 97.3%.

Keywords: dual-active bridge (DAB) converter; full-switch zero-voltage switching (ZVS); suppress the root mean square (RMS) inductor current; regulatory factor λ; full-load and ultra-voltage gain range

1. Introduction

Dual-active bridge (DAB) converters have gained widespread adoption in medium- and high-power systems, owing to their structural symmetry, inherent galvanic isolation, extended soft-switching capability, versatile power control, compact design, and ease of digital implementation. They are particularly well-suited for use in DC microgrids [1,2], automotive power systems [3,4], and energy storage systems [5].

Ensuring consistently high efficiency over a broad span of input and output voltage conditions remains a critical design target in such applications. For converters utilizing wide-bandgap (WBG) devices, achieving ZVS for all power switches is essential, as they substantially reduce switching losses—particularly at elevated switching frequencies where such losses become more pronounced [6]. Furthermore, ZVS contributes to improved electromagnetic compatibility (EMC) and system reliability by mitigating high dv/dt transients during switching events.

To extend the ZVS range and accommodate wide variations in input and output voltages, two primary approaches have been investigated, which include topological modifications [7,8] and modulation strategy enhancements. While topological changes can expand operating boundaries, they often introduce increased circuit complexity, higher cost, and reduced system reliability. As a result, modulation-based techniques are generally preferred in practical implementations due to their lower overhead and greater adaptability. DAB converters commonly employ phase-shift modulation schemes. Although certain strategies utilize asymmetric duty cycles [9–11], they often result in unbalanced switching stress and reduced transformer utilization. In contrast, symmetric duty cycle modulation offers better current sharing and improved transformer efficiency. Modulation approaches for DAB converters are generally categorized by theirs degree of freedom (DOFs), including single-phase shift (SPS) with one DOF, dual-phase shift (DPS) or extended-phase shift (EPS) with two DOFs, and triple-phase shift (TPS) offering three DOFs. Among these, SPS remains the most prevalent due to its straightforward control logic and ease of hardware realization [12]. However, it exhibits severe limitations—particularly when there is a significant deviation of the voltage conversion ratio from its nominal value—where the loss of ZVS leads to reduced efficiency [13–15]. In contrast, DPS, EPS, and TPS provide enhanced flexibility through additional control parameters. These advanced modulation schemes have been applied to various optimization objectives, including the minimization of current stress [16–29], the suppression of circulating power [30–32], and the reduction in the root mean square (RMS) current [33–37], thereby improving the overall performance of DAB converters under wide-ranging operating conditions.

An enhanced power regulation method employing EPS modulation was developed to suppress peak current, thereby achieving notable improvements in efficiency, particularly within the medium-power operating region [19]; however, its performance remains suboptimal under light-load conditions [37]. A particle swarm optimization (PSO)-based EPS scheme was developed to optimize current stress [25], and full-range ZVS with enhanced efficiency was achieved using an AI-based EPS approach in another study [38]. Nevertheless, both [25,38] rely on artificial intelligence algorithms, which suffer from limited online computation capability and poor portability, making them unsuitable for practical engineering applications. In the context of DPS modulation, ZVS conditions have been analyzed in several studies. Zhao et al. concentrated on the comprehensive reduction in different categories of power dissipation to enhance overall system efficiency [39], while Liu et al. emphasized reducing surge currents and improving system stability [40]. However, neither of these solutions achieve ZVS across the full-load range.

A current stress optimization strategy based on dual-phase shift (CSO-DPS) was proposed to alleviate current stress in both the buck and boost operation modes. However, its ZVS capability remained limited. Under conventional DPS control, at least one switch pair in the DAB converter was subjected to hard switching during high-power operation. In addition, methods leveraging magnetizing current were also explored to enable wide-range

ZVS and mitigate current stress [20,22]. While effective, these approaches introduce additional complexity in transformer design, posing challenges for practical implementation.

With its superior modulation flexibility, the TPS control enables comprehensive performance optimization over a broad operating range. A TPS-based current stress reduction approach was introduced in [16]. In [27], a reactive power controller based on TPS was innovatively introduced to reduce inductor current across all load conditions by optimizing reactive power. However, its full-range ZVS performance was not thoroughly investigated. In [41], a TPS-based strategy was proposed to minimize current stress under varying DC-link voltages. While it achieved minimal current stress at light loads, full ZVS was not guaranteed for all switches. The strategy presented in [21] realized ZVS for all six power switches under medium-load conditions and full-switch ZVS under both light and heavy loads, while maintaining nearly optimal peak inductor currents in both the buck and boost modes. Nonetheless, two switches still failed to achieve ZVS under medium load, resulting in elevated switching losses. In [23], artificial intelligence (AI)-based methods were employed to build current stress prediction models from simulation datasets. Despite its effectiveness, this approach requires extensive simulation or experimental data and a separate optimization process for each operating condition, leading to high complexity and limited real-time applicability. TPS-based current stress optimization was also investigated, but this method failed to achieve ZVS for two switches in the medium-power region, as shown in [26]. A three-level phase-shift control scheme was presented in to minimize reactive power and improve conversion efficiency across a wide operational range [30]. In addition, a unified optimal modulation strategy (UOMS) was developed to suppress circulating power, thereby achieving full-range ZVS and reduced root mean square (RMS) current. Nevertheless, the effectiveness of UOMS remains constrained to certain operating points, limiting its general applicability [32].

Given that conduction losses increase proportionally with the squared magnitude of the RMS current, minimizing RMS current is essential for improving overall energy efficiency [37]. In [28], the authors investigated the correlation between current stress and RMS current minimization under TPS control and proposed an associated optimization method. However, the manuscript does not provide a detailed analysis of ZVS, and the relatively low switching frequency raises concerns about the strategy's effectiveness at higher frequencies. While studies [33,34] have investigated RMS and efficiency optimization, their analyses were tailored to specific device configurations and lacked generalized analytical modulation expressions, limiting their applicability to broader operating conditions. A global optimal condition (GOC) for minimizing the RMS current across the entire power range was proposed, but ZVS performance was not considered in the analysis [35]. However, it neglected the evaluation of ZVS characteristics. To address RMS current, another approach grounded in fundamental harmonic analysis was proposed, yet ZVS could not be achieved by two switches under moderate and low load levels [37]. Moreover, under light-load conditions, none of the six switches achieved ZVS [16,30,35]. In [29], the authors propose a method that dynamically sets the ZVS current threshold to achieve ZVS for power switches while maintaining a low inductor current. However, under low-power conditions with high output voltage, ZVS can be achieved for only two switches, and under medium-power conditions, ZVS is limited to six switches. To improve ZVS operation and attain high efficiency across the full-load range, nonlinear modeling techniques were utilized to more accurately capture the switching behavior [42]. Nevertheless, these methods typically depend on offline lookup tables, which are unsuitable for real-time control and impose substantial computational requirements, thereby limiting their feasibility in practical implementations [33,34,42].

To address the limitations identified in prior studies, this paper proposes a symmetric optimization strategy based on triple-phase shift for DAB converters (SOS-TPS). The proposed methodology is designed to ensure full-range ZVS and reduced RMS current while maintaining high efficiency under diverse operating scenarios. First, among the eight operating modes capable of forward power transfer, three are selected for further analysis based on their optimization potential. Using the Karush–Kuhn–Tucker (KKT) conditions and the Lagrange multiplier method (LMM), the optimal combination of control variables is derived for minimizing the RMS current in the forward buck mode under low- and medium-power conditions. To facilitate full-switch ZVS at low power, a modulation factor λ is introduced, balancing soft-switching and current stress. For high-power operation, the strategy transitions to conventional SPS modulation, which inherently satisfies ZVS and maintains low RMS current. Second, exploiting the structural symmetry of the DAB topology, the proposed approach reveals a key modulation feature: once the analytical framework is established for the forward buck mode, the corresponding control law for the forward boost mode can be directly obtained via variable transformation, eliminating the need for repetitive derivation. Third, most existing modulation strategies are only effective under specific operating conditions and tend to fail when the voltage conversion ratio between the input and output varies significantly. In contrast, the proposed SOS-TPS strategy remains applicable across the full-load range and large voltage conversion ratios, maintaining both low RMS current and full-switch ZVS as its optimization objectives. Fourth, due to its closed-form analytical expressions, the proposed modulation strategy supports real-time computation and exhibits high portability, making it more suitable for practical engineering implementation. Finally, the proposed SOS-TPS strategy is experimentally verified to demonstrate its correctness and practical effectiveness.

2. Analysis Modeling and Operational Analysis of DAB Converters

2.1. Operational Mechanism of DAB Converters

As illustrated in Figure 1, the DAB converter comprises two full-bridge (FB) stages, FB_1 and FB_2, interconnected through a high-frequency transformer with a turns ratio of n:1. The voltage conversion ratio is expressed as $k = V_1/(nV_2)$, where $k > 1$ indicates operation in buck mode, and $k < 1$ corresponds to boost mode. Given the inherent structural symmetry of the DAB converter, the control strategies optimized for $k > 1$ and $k < 1$ are expected to exhibit corresponding symmetrical characteristics when V_1 remains constant.

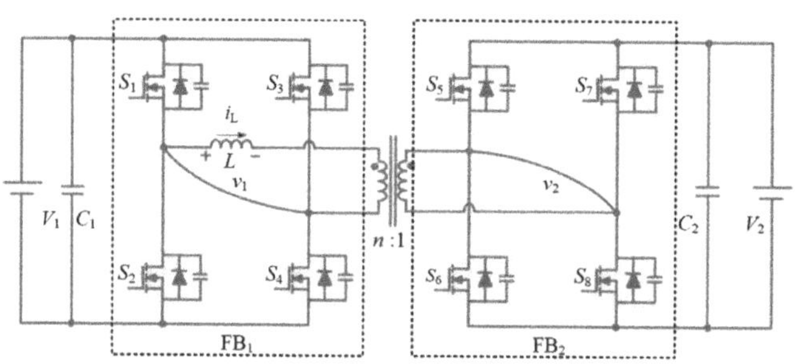

Figure 1. Topology of the DAB converter.

Figure 2 presents the representative waveforms of a DAB converter operating under the TPS modulation scheme. In this context, D_1 and D_2 denote the duty cycles of the voltage waveforms on the primary and secondary sides, respectively, while Φ signifies the phase shift between their waveform centers. These three modulation parameters—D_1, D_2, and Φ—are collectively defined as the $D_{control}$ and are restricted to the interval [0, 1]. The switching period is represented by T_s, with its associated frequency f_s, and the half-period is defined as $T_{hs} = T_s/2$. The inductor current, along with the voltages v_1 and v_2, exhibit waveform variations dependent on D_1, D_2, and Φ, which subsequently affect both the RMS value of the leakage inductor current and the average power transferred from the primary to the secondary side.

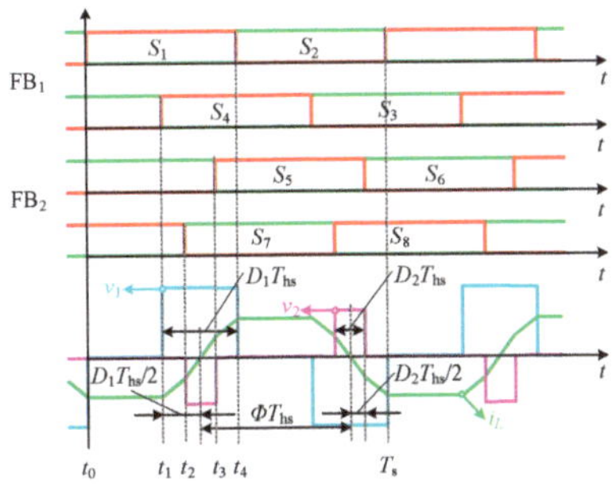

Figure 2. Triple phase-shift modulation waveforms of a DAB.

2.2. Classification and Selection of TPS Modes in DAB Converters

Depending on the specific combination of D_1, D_2, and Φ, a DAB converter can operate in eight distinct modulation modes under the TPS control scheme. Figure 3 depicts the representative voltage and current waveforms for each TPS mode. As presented in Section 2.3 of [43], the normalized transmitted power expressions, operating ranges, and mode constraints for each modulation mode can be obtained, along with the normalized inductor current values at each switching instant for modes D, F, and G.

$$I_{rms}{}^2 = \frac{2}{3T_s}\sum_{i=0}^{i=3}\left(i_L{}^2(t_i) + i_L{}^2(t_{i+1}) + i_L(t_i)i_L(t_{i+1})\right) \tag{1}$$

$$I_{nrms}{}^2 = \frac{I_{rms}{}^2}{\frac{n^2V_2{}^2}{48f_s{}^2L^2}} \tag{2}$$

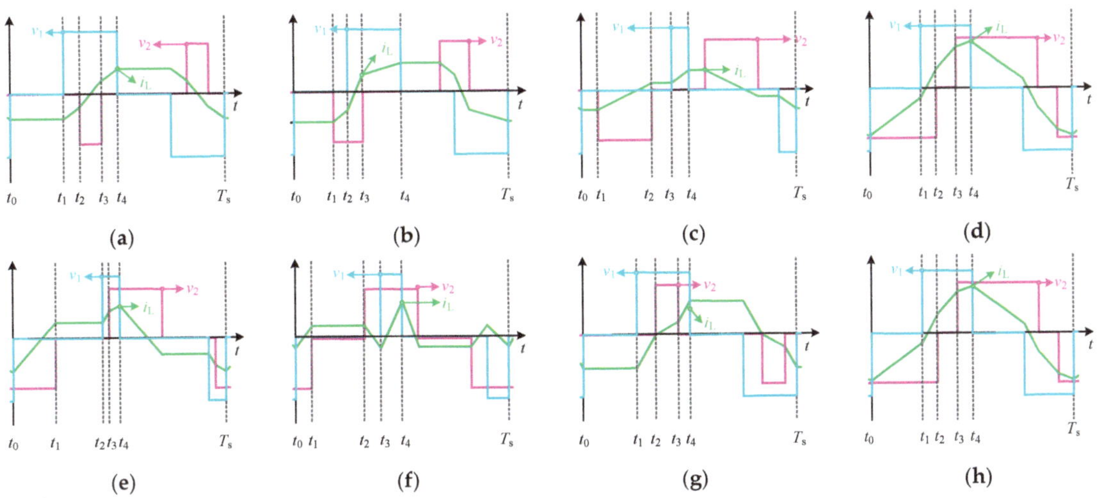

Figure 3. Typical voltage and current waveforms under eight TPS modulation modes. (**a**) Mode A. (**b**) Mode B. (**c**) Mode C. (**d**) Mode D. (**e**) Mode E. (**f**) Mode F. (**g**) Mode G. (**h**) Mode H.

This study focuses on optimizing forward power transfer of the DAB converter under both buck and boost conditions. Based on Figure 3, the normalized inductor currents for each modulation mode, and Equation (1), the actual RMS inductor current I_{rms} is computed. To facilitate further analysis, the square of the RMS current is normalized using Equation (2), and the normalized RMS current $I_{nrms}{}^2$ for each modes are summarized in Table 1.

Table 1. Normalized root mean square current values for modes D, F, and G.

Mode	Normalized Root Mean Square Current Values
D	$3D_2^2 - 2D_2^3 + 4k - 6D_1k + 3D_1^2k - 6D_2k + 3D_2^2k + 3D_1^2k^2 - 2D_1^3k^2$ $-12k\Phi + 12D_1k\Phi - 6D_1^2k\Phi + 12D_2k\Phi - 6D_2^2k\Phi + 12k\Phi^2 - 8k\Phi^3$
F	$3D_2^2 - 2D_2^3 + D_1^3k - 6D_1D_2k + 3D_1D_2^2k - 2D_1^3k^2 + 12D_1k\Phi^2$
G	${3D_2}^2 - {2D_2}^3 - 6D_1D_2k + {3D_1}^2D_2k + {D_2}^3k + {3D_1}^2k^2 - {2D_1}^3k^2 + 12D_2k\Phi^2$

2.3. Symmetry of DAB Converters

Taking mode D as an example, the DAB converter operates under condition a. By interchanging the primary and secondary DC voltages, as well as the duty cycles of the AC voltages on both sides while keeping the phase shift Φ unchanged, b is obtained. The normalized transferred power of the DAB converter depends solely on the D_{control} vector; thus, $p_a = p_b$. As shown in Figure 4, the waveforms of a and b are symmetric with respect to the red axis. The current values at the same time instants t_x ($x = 0, 1, 2, 3, 4$) are identical, indicating that the actual normalized root mean square current values under operating conditions a and b are equal, as also supported by Equation (2).

$$I_{nrms(a)}{}^2 = I_{nrms(b)}{}^2 \tag{3}$$

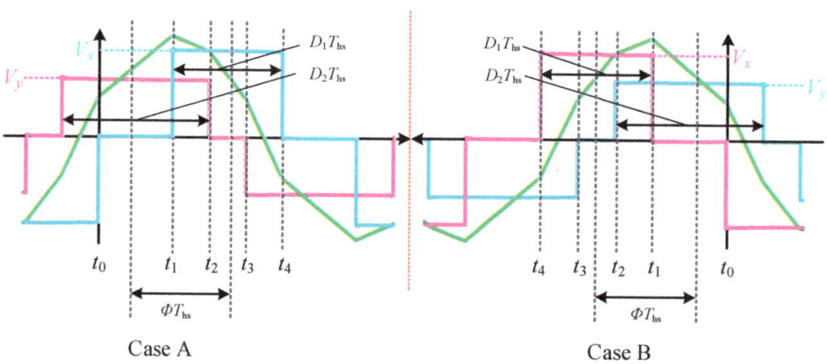

Figure 4. Operating waveforms under operating conditions a and b.

Generalizing this symmetry, consider the DAB converter operating under condition c in mode D (Table 2), where $V_1 = V_{\text{m}}$ and $V_2 = V_{\text{n}}$, with $V_{\text{m}}/V_{\text{n}}$ maintained as $1/k$. The ratio of the normalized RMS current values of operating conditions a and c can be derived as follows:

$$\frac{I_{nrms(a)}{}^2}{I_{nrms(c)}{}^2} = k^2 \tag{4}$$

Table 2. Circuit parameters under three operating conditions.

Operating Conditions	V_1	V_2	Duty Cycles of the Primary Voltage Waveforms	Duty Cycles of the Secondary-Side Voltage Waveforms
a	V_x	V_y	D_1	D_2
b	V_y	V_x	D_2	D_1
c	V_m	V_n	D_2	D_1

It is demonstrated that interchanging the DC voltages on the primary and secondary sides, swapping the duty cycles of the transformer AC voltages, and maintaining the phase shift ensures the same normalized power transfer. Moreover, the variation trends of the normalized RMS current with respect to normalized power are consistent across both cases. Based on this, the optimization strategy for forward buck mode can be directly extended to the forward boost case simply by swapping D_1 and D_2 and replacing k with $1/k$.

For mode F, let the control vector take configuration a, where D_1, D_2, and Φ satisfy the constraints of the mode. Its symmetric counterpart operates under b, which essentially corresponds to mode G. The original and symmetric configurations show similar trends in the normalized RMS current variation, with respect to transferred power.

3. Principle of Proposed SOS-TPS

In the forward buck power transfer mode, the optimization problems for modes F and D are formulated with the RMS current as the objective, subject to mode-specific constraints. First, the optimal control trajectory is obtained by minimizing the RMS current using the Karush–Kuhn–Tucker (KKT) conditions and the Lagrange multiplier method (LMM). Then, the switching current is analyzed to validate or refine the ZVS conditions, leading to the final optimized operating trajectory.

3.1. Modulation Strategy in the Low-Power Range When k > 1

Under light-load conditions in buck mode, the DAB converter operates optimally in mode F. To characterize the relationship among normalized RMS current, normalized transmitted power, and duty cycle boundaries, a Lagrange function is constructed as follows:

$$L_f = 3D_2^2 - 2D_2^3 + D_1^3 k - 6D_1 D_2 k + 3D_1 D_2^2 k - 2D_1^3 k^2 + 12D_1 k\Phi^2 + \lambda(4D_1\Phi - p) + \mu_1\left(\Phi - \frac{D_1 - D_2}{2}\right) \tag{5}$$

Here, λ serves as the Lagrange multiplier associated with the equality constraint, while μ_1 corresponds to the multiplier for the inequality constraint. By computing the partial derivatives of the Lagrange function L_f, the following expressions are derived:

$$\begin{cases} \frac{\partial L_f}{\partial D_1} = 0, \frac{\partial L_f}{\partial D_2} = 0, \frac{\partial L_f}{\partial \Phi} = 0 \\ \lambda \neq 0, \mu_1 \geq 0 \\ 4D_1\Phi - p = 0 \\ \mu_1\left(\Phi - \frac{D_1 - D_2}{2}\right) = 0 \\ \Phi - \frac{D_1 - D_2}{2} \leq 0 \\ k > 1 \end{cases} \tag{6}$$

By solving (6), the optimal $D_{control}$ can be written as:

$$\begin{cases} D_1 = \sqrt{\frac{p}{2(k-1)}} \\ D_2 = k\sqrt{\frac{p}{2(k-1)}} \\ \Phi = \frac{1}{2}\sqrt{\frac{(k-1)p}{2}} \end{cases} \quad (7)$$

It can be concluded that, for any specified normalized power p and voltage conversion ratio k, there exists a feasible combination of duty cycles D_1, D_2, and Φ that satisfies the operational constraints. The mathematical relationship among these variables is expressed as:

$$\begin{cases} \Phi = \frac{1}{2}(D_2 - D_1) \\ D_2 = kD_1 \end{cases} \quad (8)$$

According to (7), switches S_3–S_8 fail to achieve ZVS. To address this, a small amount of reactive power is introduced. For mode F, it is always true that $i_L(t_2) > i_L(t_3)$; therefore, enforcing $i_L(t_2) + i_L(t_3) = 0$ not only ensures the required polarity for ZVS at t_2 and t_3 but also provides sufficient magnitude to completely discharge the SiC MOSFETs junction capacitance. Based on the condition $i_L(t_2) + i_L(t_3) = 0$, the following expression can be derived:

$$\Phi = (k - \frac{1}{2})D_1 - \frac{1}{2}D_2 \quad (9)$$

According to (9) and the constraints of mode F, it can be deduced that $k D_1 < D_2 < 1$. To control the amplitude of $i_{Ln}(t_2)$, a modulation factor λ is introduced such that $0 < \lambda < 1$ and $D_2 = \lambda + kD_1 (1 - \lambda)$. Thus, the normalized current $i_{Ln}(t_2)$ can be rewritten as:

$$i_{Ln}(t_2) = \lambda(1 - kD_1) \quad (10)$$

The value of λ affects both $i_{Ln}(t_2)$ and $I_{nrms}{}^2$. Figure 5a,b illustrate the relationships between the normalized transferred power p and the normalized RMS current $I_{nrms}{}^2$, as well as those between p and the normalized current stress $i_{Ln}(t_2)$, respectively, for $\lambda = 0.1$, 0.3, 0.5, and 0.7. As shown in Figure 5, for a given transferred power, both $I_{nrms}{}^2$ and $i_{Ln}(t_2)$ increase with increasing λ. In practice, a larger $i_{Ln}(t_2)$ facilitates the realization of ZVS but leads to a higher $I_{nrms}{}^2$. Therefore, a trade-off in selecting λ is necessary. In this study, $\lambda = 0.5$ is adopted.

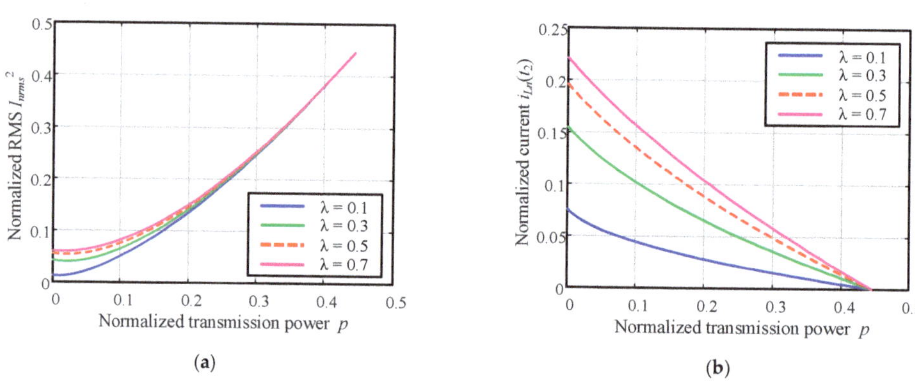

Figure 5. Influence of λ on the normalized RMS current $I_{nrms}{}^2$ and the normalized inductor current $i_{Ln}(t_2)$. (**a**) Variation of $I_{nrms}{}^2$ with normalized transmitted power p. (**b**) Variation of $i_{Ln}(t_2)$ with normalized transmitted power p.

The normalized transferred power range of mode F is [0, $2(k-1)/k^2$]. The final optimal $D_{control}$ for mode F is given by:

$$\begin{cases} D_1 = \frac{\sqrt{1-8p+12kp}+1}{2(3k-2)} \\ D_2 = \frac{1}{2}(1+D_1k) \\ \Phi = \frac{1}{4}(-1+(-2+3k)D_1) \end{cases} \tag{11}$$

3.2. Modulation Strategy in Medium-Power Range When k > 1

In the buck mode operation of the DAB converter, mode D is selected as the optimal solution. The Lagrange function is formulated to describe the relationship between the normalized RMS current $I_{nrms}{}^2$, normalized transmitted power p, and duty cycle constraints within the selected switching mode as follows:

$$\begin{aligned} L_f = {} & 3D_2{}^2 - 2D_2{}^3 + 4k - 6D_1k + 3D_1{}^2k - 6D_2k + 3D_2{}^2k \\ & + 3D_1{}^2k^2 - 2D_1{}^3k^2 - 12k\Phi + 12D_1k\Phi - 6D_1{}^2k\Phi + 12D_2k\Phi - 6D_2{}^2k\Phi + 12k\Phi^2 - 8k\Phi^3 \\ & + \lambda\left(1-(1-D_1)^2-(1-D_2)^2-(1-2\Phi)^2-p\right) + \mu_1\left(1-\frac{D_1+D_2}{2}-\Phi\right) + \mu_2\left(\Phi-\left(\frac{D_1+D_2}{2}\right)\right) \end{aligned} \tag{12}$$

Here, λ represents the weight of the equality constraint, while μ_1 and μ_2 denote the weights of the inequality constraints. By taking the partial derivative of the Lagrange function L_f, the following results can be obtained:

$$\begin{cases} \frac{\partial L_f}{\partial D_1} = 0, \frac{\partial L_f}{\partial D_2} = 0, \frac{\partial L_f}{\partial \Phi} = 0 \\ \lambda \neq 0, \mu_1, \mu_2 \geq 0 \\ 1-(1-D_1)^2-(1-D_2)^2-(1-2\Phi)^2-p = 0 \\ \mu_1\left(1-\frac{D_1+D_2}{2}-\Phi\right) = 0 \\ \mu_2\left(\Phi-\left(\frac{D_1+D_2}{2}\right)\right) = 0 \\ k > 1 \end{cases} \tag{13}$$

By jointly analyzing (12) and (13), the optimal combination is determined as follows:

$$\begin{cases} D_1 = (1-k+2k\Phi) + \sqrt{(k^2+1)(1-2\Phi)^2 - 2k(1-2\Phi)+1} \\ D_2 = 1 \end{cases} \tag{14}$$

When the voltage conversion ratio $k > 1$, mode D inherently satisfies the current conditions $i_L(t_0) > 0$ and $i_L(t_2) < i_L(t_3)$ throughout its operation. Therefore, the full ZVS conditions for mode D are designated as $i_L(t_1) < 0$ and $i_L(t_2) > 0$. By incorporating the ZVS constraints and mode boundary conditions for mode D, the following relation is derived:

$$\begin{cases} \Phi > \frac{1}{2}(2-D_1-D_1k) & i_{Ln}(t_1) < 0 \\ \Phi > \frac{-D_2+2k-D_2k}{2k} & i_{Ln}(t_2) > 0 \\ \frac{1-D_1}{2} < \Phi < \frac{1+D_1}{2} \\ p = 1-(1-D_1)^2-(1-2\Phi)^2 \end{cases} \tag{15}$$

Based on (14), a functional relationship among the normalized transferred power p, duty cycle D_1, and phase shift Φ in mode D can be established. Taking k = 1.5 as an example, the three-dimensional surface and its two-dimensional projection are depicted in Figure 6. Additionally, as shown in Figure 6, the feasible region of the control variables (D_1, Φ) is bounded by the quadrilateral DCIH. The optimal control trajectory remains within this region and gradually approaches the red zone as the transferred power increases. The optimal control trajectory lies entirely within this region and gradually approaches the red area as power increases. According to (14) and (15), the normalized power range for mode D is given by $[2(k-1)/k^2, 2(1-k^2+k(k^2-1)^{1/2})]$, point L corresponds to the minimum normalized transmitted power $p_L = 2(k-1)/k^2$, while point M represents the maximum, given by $p_M = 2(1-k^2+k(k^2-1)^{1/2})$.

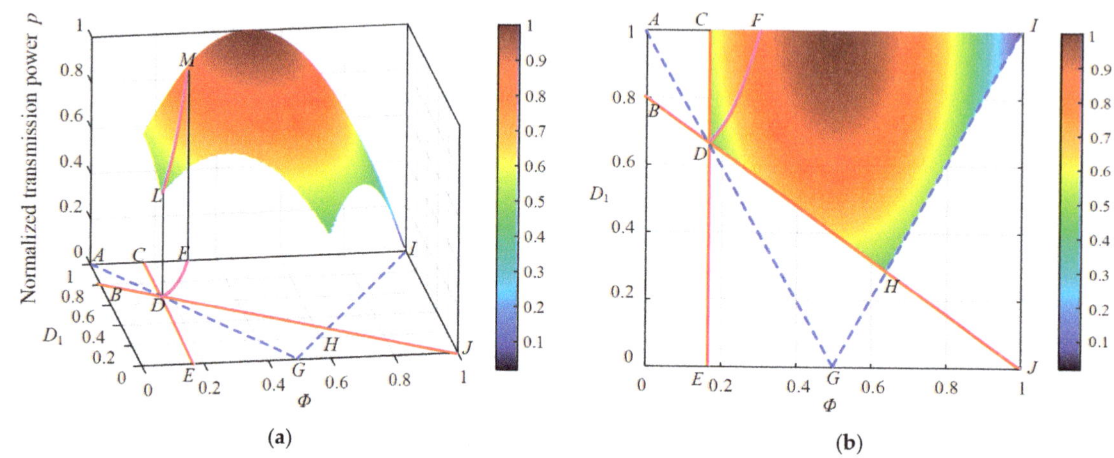

Figure 6. Optimized operating path in the medium region for $k > 1$. (**a**) Three-dimensional visualization of the optimized path. (**b**) Two-dimensional projection of the optimized path.

The above analysis confirms that for $k = 1.5$, the optimal trajectory of mode D remains within its ZVS constraints and feasible operating domain. To further investigate the applicability of this trajectory for other $k > 1$ conditions, the polarity relationship between $i_{Ln}(t_1)$ and $i_{Ln}(t_2)$ under different values of k and normalized power p is analyzed based on mode D constraints and Equation (14), as shown in Figure 7.

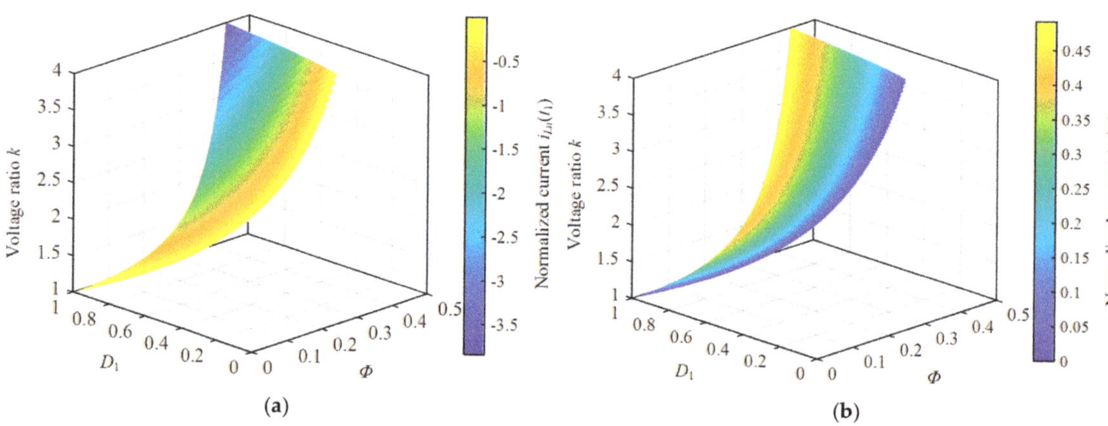

Figure 7. Variation of $i_{Ln}(t_1)$ and $i_{Ln}(t_2)$ with respect to D_1, Φ and k. (**a**) $i_{Ln}(t_1)$. (**b**) $i_{Ln}(t_2)$.

Figure 7 illustrates the feasible solution space of the $D_1 - \Phi$ domain that satisfies both the mode D operational constraints and the condition given in Equation (13). As shown in the 3D color maps, the conditions $i_{Ln}(t_1) < 0$ and $i_{Ln}(t_2) > 0$ are consistently satisfied, indicating that the derived optimal trajectory enables ZVS for all switches. It is evident from Figure 7a,b that this polarity relation holds within the range of $1 < k < 4$. In fact, by combining Equation (14) with the mode D operational constraints, it can be generally proven that the optimal trajectory ensures full-switch ZVS for any $k > 1$.

3.3. Modulation Strategy in the High-Power Range When $k > 1$

Under high-power conditions, the DAB converter operates with conventional single-phase shift (SPS) modulation, and the control variable combination is defined as follows:

$$\begin{cases} D_1 = 1 \\ D_2 = 1 \end{cases} \tag{16}$$

Figure 8 illustrates the variation in control parameters D_1, D_2, and Φ with respect to the normalized transmitted power p. When $0 < p < 2(k - 1)/k^2$, the converter operates in mode F, and all three parameters (D_1, D_2, and Φ) increase monotonically with p. At $p = 2(k - 1)/k^2$, D_2 reaches one and remains constant thereafter. As p increases further within the range $2(k - 1)/k^2 < p < 2(1 - k^2 + k(k^2 - 1)^{1/2})$, the system transitions into mode D, where D_2 and Φ continue to increase. When $p = 2(1 - k^2 + k(k^2 - 1)^{1/2})$, D_1 reaches one and the control scheme evolves into SPS modulation, with the transmitted power solely increasing with Φ. Throughout the entire process, variations in D_1, D_2, and Φ are continuous without abrupt changes, ensuring smooth transitions between operating modes.

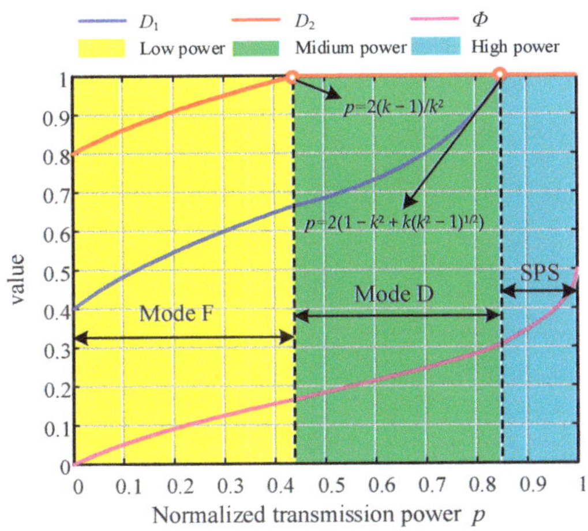

Figure 8. Variation in D_1, D_2, and Φ with normalized transmission power p.

3.4. Synopsis of the SOS-TPS Strategy for DAB Converters

As concluded in Section 2.3, the modulation strategy for the forward boost mode can be systematically derived from its buck-mode counterpart by exploiting the inherent structural symmetry of the DAB converter. A summary of the derived results is provided in Table 3.

Table 3. The SOS-TPS modulation strategy across the full-power range in forward buck and boost modes.

Modes	Low-Power Range	Medium-Power Range	High-Power Range
$k>1$	$0<p<\frac{2(-1+k)}{k^2}$	$\frac{2(-1+k)}{k^2}<p<2\left(1-k^2+k\sqrt{k^2-1}\right)$	$2\left(1-k^2+k\sqrt{k^2-1}\right)<p<1$
	$\begin{cases} D_1=\frac{\sqrt{1-8p+12kp}+1}{2(3k-2)} \\ D_2=\frac{1}{2}(1+D_1k) \\ \Phi=\frac{1}{4}(-1+(-2+3k)D_1) \end{cases}$	$\begin{cases} D_1=1-k+2k\Phi+\sqrt{(k^2+1)(1-2\Phi)^2-2k(1-2\Phi)+1} \\ D_2=1 \end{cases}$	$\begin{cases} D_1=1 \\ D_2=1 \end{cases}$
$k<1$	$0<p<2k(1-k)$	$2k(1-k)<p<\frac{2\left(-1+k^2+\sqrt{1-k^2}\right)}{k^2}$	$\frac{2\left(-1+k^2+\sqrt{1-k^2}\right)}{k^2}<p<1$
	$\begin{cases} D_1=\frac{1}{2}\left(1+\frac{D_2}{k}\right) \\ D_2=\frac{k\left(1+\sqrt{1+\left(-8+\frac{12}{k}\right)p}\right)}{6-4k} \\ \Phi=\frac{1}{4}\left(-1+\left(-2+\frac{3}{k}\right)D_2\right) \end{cases}$	$\begin{cases} D_1=1 \\ D_2=\frac{-1+k+2\Phi+\sqrt{k^2(1-2\Phi)^2+(-1+k+2\Phi)^2}}{k} \end{cases}$	$\begin{cases} D_1=1 \\ D_2=1 \end{cases}$

4. Comparative Analysis of ZVS Range and RMS Current in Existing Schemes

Due to the intrinsic symmetry of the DAB converter, the buck and boost modes exhibit similar characteristics under forward power transfer. Therefore, this section provides a comparative analysis of the ZVS range and RMS current performance in buck mode for the SPS [12], GOC [35], UOMS [32] and the proposed SOS-TPS modulation strategies.

4.1. ZVS Range

Figure 9 depicts the two-dimensional operating regions of normalized power p versus the voltage conversion ratio k under SPS, GOC, and UOMS modulation strategies. The green curve marks the transition boundary between partial and full ZVS under SPS modulation. Full-switch ZVS is only achievable when p lies above the curve; in the low-power region, only four switches can achieve ZVS.

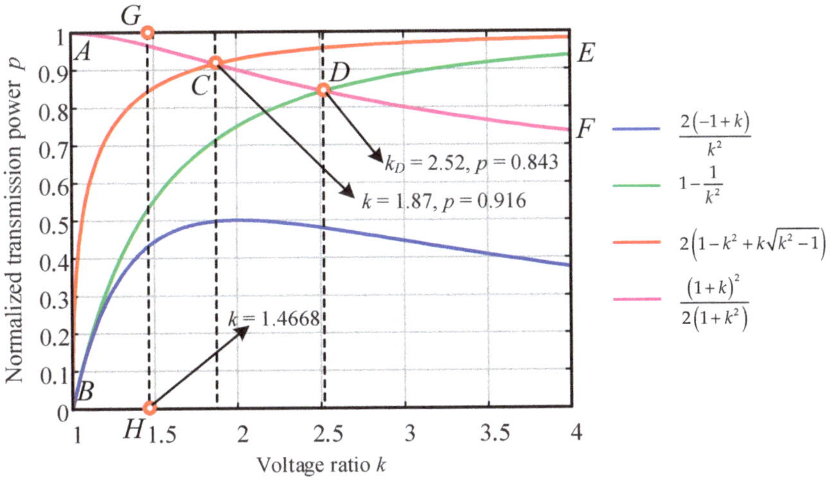

Figure 9. ZVS region mapping of various modulation strategies.

The blue and red curves jointly define the power regions under GOC modulation. Although the area below the blue curve achieves RMS current minimization, all six switches operate in ZCS, leading to significant turn-on losses under high-frequency conditions, thus limiting efficiency improvement.

The blue and magenta curves define the power regions of the UOMS modulation strategy. Experimental validation on a prototype with $k = 1.4668$ (operating along line *GH*) demonstrates that full-switch ZVS can be achieved across the entire load range while maintaining a low RMS current. However, theoretical analysis indicates that when $k > 2.52$, the DAB converter should transition to SPS in the high-power region. Since SPS cannot achieve full-switch ZVS in region DEF, the applicability of the UOMS modulation strategy is limited to $k < k_D$.

Here, the blue curve defines the low-power boundary. Using the SOS-TPS modulation strategy, full-switch ZVS is achieved in the low-power region while maintaining a low RMS current. In the medium-power region below the red curve, the proposed strategy preserves the RMS current minimization characteristics. For the high-power region, the control naturally transitions to SPS modulation. Overall, the proposed method theoretically enables full-switch ZVS and low RMS current across the entire load range for any $k > 1$, overcoming the efficiency and applicability limitations of existing modulation strategies.

4.2. RMS Current

A comparative evaluation of the SPS, GOC, Unified, and proposed SOS-TPS modulation strategies is conducted under identical operating conditions to verify the efficiency improvement achieved by the proposed method. In this section, the RMS current is calculated using MATLAB R2020a based on the parameters listed in Table 4.

Table 4. Design specifications of the DAB converter experimental platform.

Items	Specifications
Operating Condition Configuration	Input port: $V_1 = 80$ V Output port: $V_2 = 26.66$ V, 53.33 V, 120 V, 240 V
Switching Frequency	$F_s = 40$ kHz
Switches	S_1–S_8: C3M0075120K $V_{DSS} = 1200$ V, $I_D = 32$ A, $R_{ds(on)} = 75$ mΩ, Coss = 58 pF
Inductor	Inductance: 25.5 μH
Transformer	Turns Ratio $n = 1$
Controller	TMS320F28335
Filter Capacitor	$C_1 = C_2 = 600$ μF

Figure 10 illustrates the variation in the normalized RMS current $I_{nrms}{}^2$ with respect to the normalized transferred power p. In the range $0 < p < 0.3$, the $I_{nrms}{}^2$ values of the SOS-TPS and Unified strategies are slightly higher than that of GOC, primarily due to the introduction of a small amount of reactive power to achieve full-switch ZVS. However, both are significantly lower than that of SPS, which maintains a relatively high $I_{nrms}{}^2$ throughout the range $0 < p < 0.65$. In the range $0.65 < p < 0.75$, all four strategies yield comparable RMS current levels. Between $0.75 < p < 0.96$, the Unified scheme exhibits a noticeably higher $I_{nrms}{}^2$ than the other three, which remain nearly identical. At $p = 0.96$, the UOMS transitions to SPS. For $0.96 < p < 1$, all four strategies converge to the same RMS current level as they operate in the SPS mode.

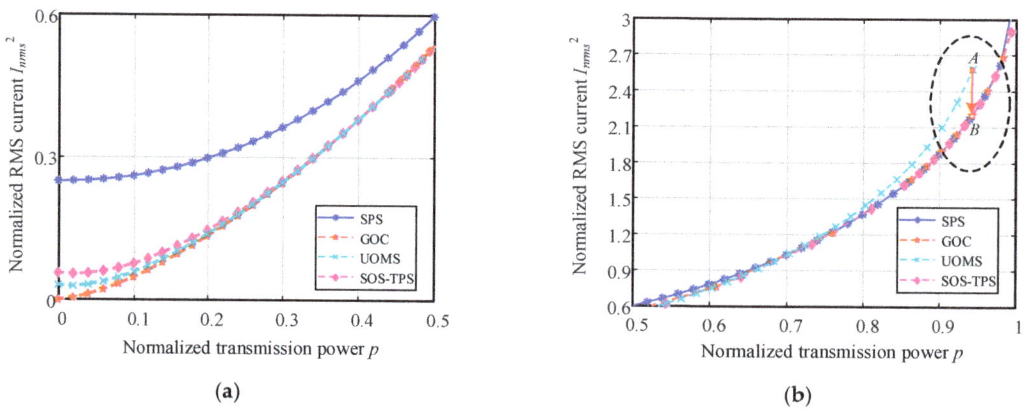

Figure 10. Normalized RMS current $I_{nrms}{}^2$ versus normalized transferred power p. (**a**) $0 < p < 0.5$. (**b**) $0.5 < p < 1$.

5. Experimental Verification

To validate the feasibility of the proposed control strategy, as illustrated in Figure 11, an experimental platform based on a DAB converter is established. This section first presents typical operating waveforms and ZVS performance of the SOS-TPS strategy under both buck and boost modes across different power levels. Then, the applicability of SOS-TPS under a wide voltage conversion ratio is demonstrated. Finally, the RMS current and efficiency across the full-power range are compared for SPS, GOC, UOMS, and the proposed SOS-TPS under the forward buck operating mode. According to the aforementioned symmetry characteristics, the DAB converter exhibits similar trends in RMS current and efficiency under the forward boost mode.

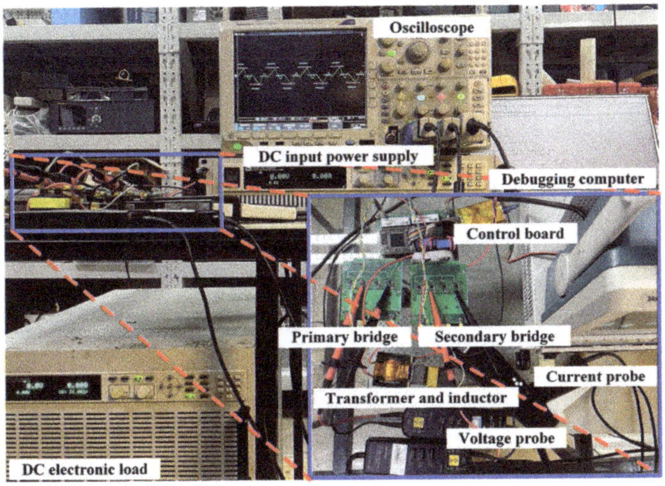

Figure 11. Experimental setup of the DAB platform.

5.1. Experimental Verification with $k > 1$

Figures 12–14 illustrate the experimental waveforms of the proposed SOS-TPS modulation strategy with a voltage conversion ratio of $k = 1.5$ ($V_1 = 80$ V, $V_2 = 53.33$ V). The results demonstrate a high degree of consistency between the measured transformer voltage and inductor current waveforms and the theoretical predictions when the DAB converter operates in mode F at 78 W, mode D at 392 W, and SPS modulation at 470 W.

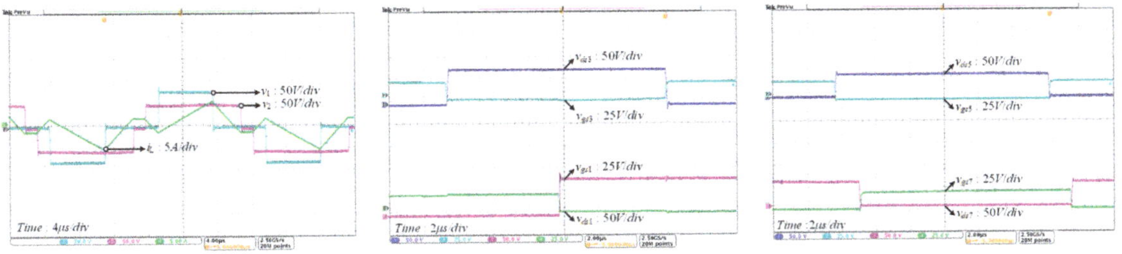

(a) (b) (c)

Figure 12. Experimental waveforms at low power (78 W) using SOS-TPS ($V_1 = 80$ V, $V_2 = 53.33$ V). (**a**) The waveforms of the voltages and current during operation. (**b**) Switching waveforms of S_1 and S_3. (**c**) Switching waveforms of S_5 and S_7.

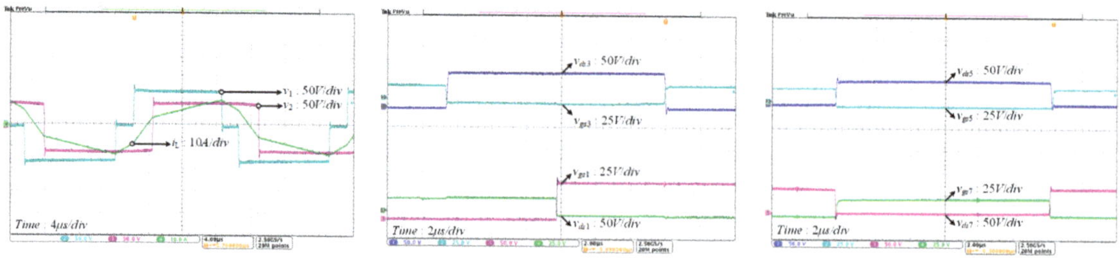

(a) (b) (c)

Figure 13. Experimental waveforms at medium power (392 W) using SOS-TPS (V_1 = 80 V, V_2 = 53.33 V). (**a**) The waveforms of the voltages and current during operation. (**b**) Switching waveforms of S_1 and S_3. (**c**) Switching waveforms of S_5 and S_7.

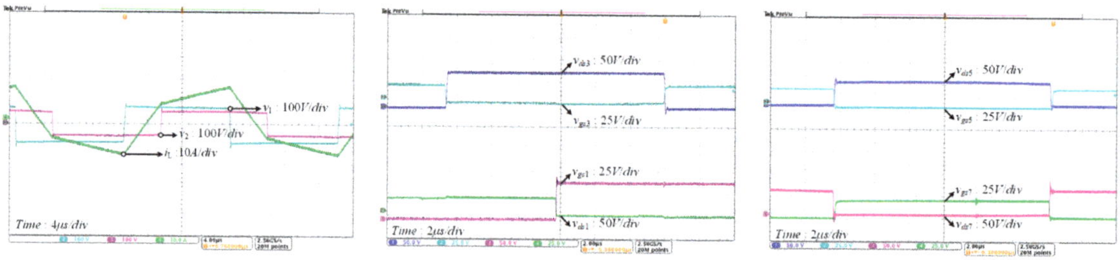

(a) (b) (c)

Figure 14. Experimental waveforms at high power (470 W) using SOS-TPS (V_1 = 80 V, V_2 = 53.33 V). (**a**) The waveforms of the voltages and current during operation. (**b**) Switching waveforms of S_1 and S_3. (**c**) Switching waveforms of S_5 and S_7.

Figure 15 presents the experimental waveform of the proposed SOS-TPS modulation strategy with a voltage conversion ratio of $k = 3$ (V_1 = 80 V, V_2 = 26.66 V). When operating in mode D at an output power of 221 W, the DAB converter demonstrates successful ZVS across all switches. These results confirm that the SOS-TPS modulation strategy maintains full-switch ZVS capability even under high voltage conversion ratio conditions.

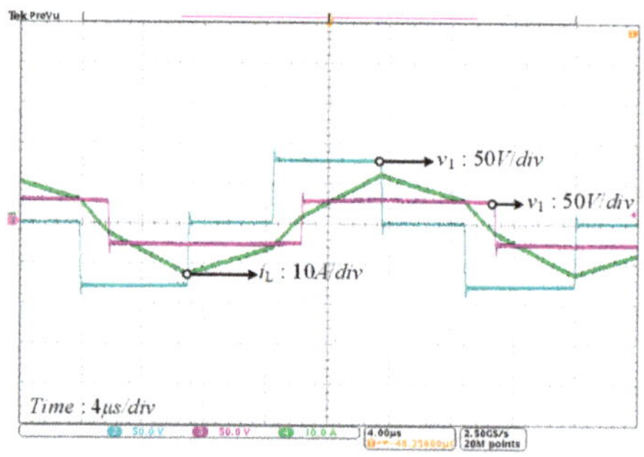

Figure 15. Experimental waveforms at medium power (221 W) using SOS-TPS (V_1 = 80 V, V_2 = 26.66 V).

5.2. Experimental Verification with k < 1

The experimental waveforms of the proposed SOS-TPS modulation strategy with a voltage conversion ratio of $k = 0.667$ (V_1 = 80 V, V_2 = 120 V) are shown in Figure 16. The results demonstrate that when the converter operates in mode G with an output power of

176 W, mode D with 823 W, and in SPS mode with 1058 W, the transformer voltage and inductor current waveforms closely match the theoretical analysis. Moreover, all switches of the DAB converter successfully achieve ZVS.

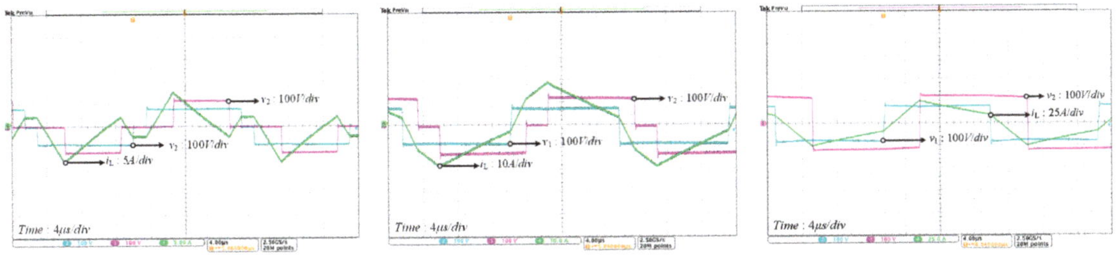

(**a**) (**b**) (**c**)

Figure 16. Experimental waveforms using SOS-TPS (V_1 = 80 V, V_2 = 120 V). (**a**) Mode G (176 W). (**b**) Mode D (823 W). (**c**) SPS (1058 W).

The experimental waveforms of the proposed SOS-TPS modulation strategy with a voltage conversion ratio of $k = 1/3$ (V_1 = 80 V, V_2 = 240 V) are shown in Figure 17. When the DAB converter operates in mode D with an output power of 2000 W, all switches successfully achieve ZVS, validating the capability of the SOS-TPS strategy to realize full-switch ZVS even under low-k conditions.

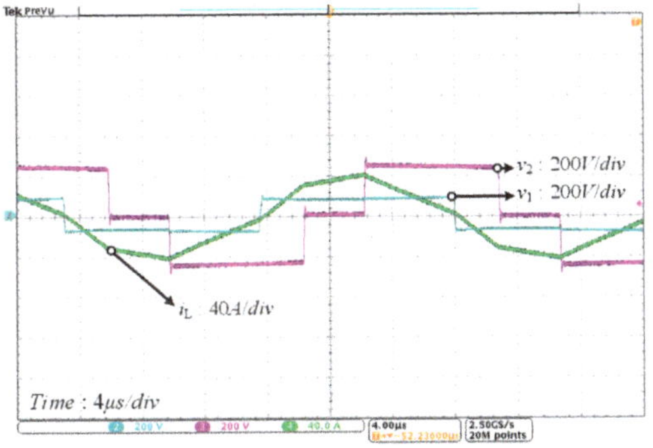

Figure 17. Experimental waveforms at medium power (2000 W) using SOS-TPS (V_1 = 80 V, V_2 = 240 V).

5.3. RMS Current and Efficiency

As illustrated in Figure 18a, the RMS current increases with output power p, exhibiting good agreement with the theoretical analysis. Figure 18b presents the measured efficiency across different output power levels. In the low-power region, UOMS and SOS-TPS achieve noticeably higher efficiency compared to GOC and SPS, with GOC slightly outperforming SPS, consistent with theoretical predictions. In the medium-power region, UOMS, GOC, and SOS-TPS maintain marginally higher efficiency than SPS. However, due to the relatively higher RMS current, the efficiency of UOMS slightly decreases in the medium-to-high power range. Near full-load, all strategies converge to SPS modulation, resulting in nearly identical efficiency. In summary, the proposed SOS-TPS enables full-range ZVS and a consistently low RMS current, thereby achieving high efficiency across the entire load range.

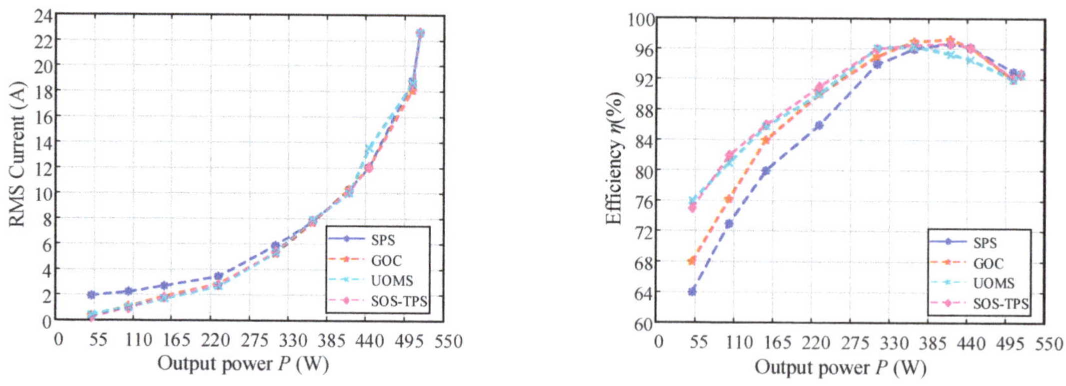

(a) (b)

Figure 18. RMS current and efficiency versus output power p. (**a**) RMS current versus output power p. (**b**) Efficiency versus output power p.

6. Conclusions

To address the efficiency limitations of traditional triple-phase shift (TPS) control with ultra-wide voltage conversion ratios and full-load conditions, a symmetric optimized strategy for TPS (SOS-TPS) is proposed based on the structural symmetry of the dual-active bridge (DAB) topology. By employing the Karush–Kuhn–Tucker (KKT) conditions and the Lagrange multiplier method, optimal control trajectories are derived for different power regions: in the low-power region, a modulation factor is introduced to achieve both a low RMS current and full-switch ZVS; in the medium-power region, RMS current minimization is adopted, inherently ensuring ZVS; and in the high-power region, the control naturally transitions to single phase-shift (SPS) modulation while maintaining full-switch ZVS and a low RMS current. Leveraging the intrinsic symmetry of the DAB converter, the control strategy for the forward boost mode can be directly obtained from the buck mode via duty ratio and voltage ratio mapping without additional analysis. Compared to existing modulation schemes, SOS-TPS enables full-switch ZVS and a low RMS current across a wider voltage conversion range and full-load conditions. By jointly optimizing switching and conduction losses, the proposed strategy ensures high efficiency throughout the entire operating range, offering an effective solution for enhancing the performance of high-power-density power electronic systems.

Author Contributions: Conceptualization, L.C.; methodology, L.C.; visualization, L.C.; project administration, Y.Z.; validation, X.W.; writing—review and editing, D.Z. All authors have read and agreed to the published version of the manuscript.

Funding: This study is supported by the National Key R&D Program of China (Grant No. 2022YFF0706202).

Data Availability Statement: The original contributions presented in the study are included in the article; further inquiries can be directed to the corresponding author.

Acknowledgments: All authors thank the editors, referees, and officers of *Electronics* for their valuable suggestions and help.

Conflicts of Interest: The authors declare no conflicts of interest.

Abbreviations

The following abbreviations are used in this manuscript:

DAB	Dual-active bridge (converter)
TPS	Triple-phase shift modulation
SOS-TPS	Symmetric optimization strategy based on triple-phase shift
RMS	Root mean square
ZVS	Zero-voltage switching
KKT	Karush–Kuhn–Tucker (conditions)
LMM	Lagrange multiplier method
SPS	Single-phase shift modulation
DPS	Dual-phase shift modulation
EPS	Extended-phase shift modulation
DOFs	Degree of freedoms
WBG	Wide bandgap
EMC	Electromagnetic compatibility
UOMS	Unified optimal modulation strategy
GOC	Global optimal condition
RL	Reinforcement learning
ANN	Artificial neural network
AI	Artificial intelligence
PSO	Particle swarm optimization
FIS	Fuzzy inference system

References

1. Hou, N.; Li, Y.W. A Tunable Power Sharing Control Scheme for the Output-Series DAB DC-DC System With Independent or Common Input Terminals. *IEEE Trans. Power Electron.* **2019**, *34*, 9386–9391. [CrossRef]
2. Tian, J.; Zhuo, C.; Wang, F.; Deng, H. Dual-Side Asymmetric Duty Modulation Based on Accurate Soft-Switching Characteristics Modeling for DAB-Based DC Microgrid. *IEEE J. Emerg. Sel. Top. Power Electron.* **2024**, *12*, 3146–3160. [CrossRef]
3. Gou, Y.; Zhuo, F.; Wang, F.; Yu, K.; Zhan, C.; Xia, B. Lifetime Estimation of SiC Power Modules in DAB Converter Over Wide-Load Range Considering the Temperature-Dependent Characteristics. *IEEE J. Emerg. Sel. Top. Power Electron.* **2024**, *12*, 156–171. [CrossRef]
4. Tian, J.; Zhuo, C.; Wang, F.; Deng, H. An RMS Current Minimization Method for Three-Level ANPC-DAB-Based Distributed Energy Storage System With Full Operation ZVS. *IEEE J. Emerg. Sel. Top. Power Electron.* **2024**, *12*, 2388–2405. [CrossRef]
5. Li, C.; Liu, J.; Du, S.; Dang, H.; Deng, Z.; Chen, H. A T-Type DAB-Based Isolated DC-DC-AC Three-Port Converter With High Power Efficiency. *IEEE Trans. Power Electron.* **2023**, *38*, 14178–14194. [CrossRef]
6. Shah, S.S.; Iyer, V.M.; Bhattacharya, S. Exact Solution of ZVS Boundaries and AC-Port Currents in Dual Active Bridge Type DC-DC Converters. *IEEE Trans. Power Electron.* **2019**, *34*, 5043–5047. [CrossRef]
7. Pan, X.; Rathore, A.K. Novel Bidirectional Snubberless Naturally Commutated Soft-Switching Current-Fed Full-Bridge Isolated DC/DC Converter for Fuel Cell Vehicles. *IEEE Trans. Ind. Electron.* **2014**, *61*, 2307–2315.
8. Xu, G.; Sha, D.; Xu, Y.; Liao, X. Hybrid-Bridge-Based DAB Converter With Voltage Match Control for Wide Voltage Conversion Gain Application. *IEEE Trans. Power Electron.* **2018**, *33*, 1378–1388. [CrossRef]
9. Xu, D.H.; Zhao, C.H.; Fan, H.F. A PWM plus phase-shift control bidirectional DC-DC converter. *IEEE Trans. Power Electron.* **2004**, *19*, 666–675. [CrossRef]
10. Mahdavifard, M.; Mazloum, N.; Zahin, F.; KhakparvarYazdi, A.; Abasian, A.; Khajehoddin, S.A. An Asymmetrical DAB Converter Modulation and Control Systems to Extend the ZVS Range and Improve Efficiency. *IEEE Trans. Power Electron.* **2022**, *37*, 12774–12792. [CrossRef]
11. Wang, M.; Wei, S.; Mou, D.; Wu, P. Research on Efficient Single-Sided Asymmetric Modulation Strategy for Dual Active Bridge Converters in Wide Voltage Range. *IEEE J. Emerg. Sel. Top. Power Electron.* **2022**, *10*, 5738–5748. [CrossRef]
12. Krismer, F.; Kolar, J.W. Accurate Small-Signal Model for the Digital Control of an Automotive Bidirectional Dual Active Bridge. *IEEE Trans. Power Electron.* **2009**, *24*, 2756–2768. [CrossRef]
13. Bai, H.; Mi, C. Eliminate Reactive Power and Increase System Efficiency of Isolated Bidirectional Dual-Active-Bridge DC-DC Converters Using Novel Dual-Phase-Shift Control. *IEEE Trans. Power Electron.* **2008**, *23*, 2905–2914. [CrossRef]

14. Zhao, B.; Yu, Q.; Sun, W. Extended-Phase-Shift Control of Isolated Bidirectional DC-DC Converter for Power Distribution in Microgrid. *IEEE Trans. Power Electron.* **2012**, *27*, 4667–4680. [CrossRef]
15. Yu, H.; Hang, L.; Zheng, X.; He, Z.; He, Y.; Shen, L.; Shao, C.; Zeng, P.; Wu, Q.; Yang, X.; et al. Globally Unified ZVS and Quasi-Optimal Minimum Conduction Loss Modulation of DAB Converters. *IEEE Trans. Transp. Electrif.* **2022**, *8*, 3989–4000. [CrossRef]
16. Hou, N.; Song, W.S.; Wu, M.Y. Minimum-Current-Stress Scheme of Dual Active Bridge DC-DC Converter With Unified Phase-Shift Control. *IEEE Trans. Power Electron.* **2016**, *31*, 8552–8561. [CrossRef]
17. Huang, J.; Wang, Y.; Li, Z.; Lei, W. Unified Triple-Phase-Shift Control to Minimize Current Stress and Achieve Full Soft-Switching of Isolated Bidirectional DC-DC Converter. *IEEE Trans. Ind. Electron.* **2016**, *63*, 4169–4179. [CrossRef]
18. Gu, Q.; Yuan, L.; Nie, J.; Sun, J.; Zhao, Z. Current Stress Minimization of Dual-Active-Bridge DC-DC Converter Within the Whole Operating Range. *IEEE J. Emerg. Sel. Top. Power Electron.* **2019**, *7*, 129–142. [CrossRef]
19. An, F.; Song, W.; Yang, K.; Luo, S.; Feng, X. Optimised power control and balance scheme for the output parallel dual-active-bridge DC-DC converters in power electronic traction transformer. *IET Power Electron.* **2019**, *12*, 2295–2303. [CrossRef]
20. Li, L.; Xu, G.; Xiong, W.; Liu, D.; Su, M. An Optimized DPS Control for Dual-Active-Bridge Converters to Secure Full-Load-Range ZVS With Low Current Stress. *IEEE Trans. Transp. Electrif.* **2022**, *8*, 1389–1400. [CrossRef]
21. Li, J.; Luo, Q.M.; Mou, D.; Wei, Y.Q.; Zhang, X.Y. Comprehensive Optimization Modulation Scheme of Low Current Level and Wide ZVS Range for Dual Active Bridge Converter With Dead-Zone Control. *IEEE Trans. Power Electron.* **2022**, *37*, 2731–2748. [CrossRef]
22. Chaurasiya, S.; Singh, B. A Load Adaptive Hybrid DPS Control for DAB to Secure Minimum Current Stress and Full ZVS Operation Over Wide Load and Voltage Conversion Ratio. *IEEE Trans. Ind. Appl.* **2023**, *59*, 1901–1911. [CrossRef]
23. Li, X.; Zhang, X.; Lin, F.; Sun, C.; Mao, K. Artificial-Intelligence-Based Triple Phase Shift Modulation for Dual Active Bridge Converter With Minimized Current Stress. *IEEE J. Emerg. Sel. Top. Power Electron.* **2023**, *11*, 4430–4441. [CrossRef]
24. Tian, J.; Wang, F.; Zhuo, F.; Cui, X.; Yang, D. An Optimal Primary-Side Duty Modulation Scheme With Minimum Peak-to-Peak Current Stress for DAB-Based EV Applications. *IEEE Trans. Ind. Electron.* **2023**, *70*, 6798–6808. [CrossRef]
25. Jia, D.; Wang, D. Current Stress Minimization Based on Particle Swarm Optimization for Dual Active Bridge DC-DC Converter. *Actuators* **2024**, *13*, 421. [CrossRef]
26. Cheng, S.; Wang, F.; Han, Z.; Zhang, C. DAB unified ZVS control strategy with optimal current stress in full power range under TPS control. *IET Power Electron.* **2024**, *17*, 2405–2415. [CrossRef]
27. Naseem, H.; Seok, J.-K. Reactive Power Controller for Single Phase Dual Active Bridge DC–DC Converters. *IEEE Access* **2023**, *11*, 141537–141546. [CrossRef]
28. Zhang, L.; Tu, C.; Xiao, F.; Liu, B.; Li, P. Intrinsic Regularity Analysis and Optimization Control of Current Stress and RMS for Dual-Active Bridge Converter. *Electronics* **2024**, *13*, 4802. [CrossRef]
29. Gong, L.; Jin, X.; Xu, J.; Deng, Z.; Li, H.; Soeiro, T.B.; Wang, Y. A Dynamic ZVS-Guaranteed and Seamless-Mode-Transition Modulation Scheme for the DAB Converter That Maximizes the ZVS Range and Lowers the Inductor RMS Current. *IEEE Trans. Power Electron.* **2022**, *37*, 13119–13134. [CrossRef]
30. Shi, H.; Wen, H.; Chen, J.; Hu, Y.; Jiang, L.; Chen, G. Minimum-Reactive-Power Scheme of Dual-Active-Bridge DC-DC Converter With Three-Level Modulated Phase-Shift Control. *IEEE Trans. Ind. Appl.* **2017**, *53*, 5573–5586. [CrossRef]
31. Shi, H.; Wen, H.; Chen, J.; Hu, Y.; Jiang, L.; Chen, G.; Ma, J. Minimum-Backflow-Power Scheme of DAB-Based Solid-State Transformer With Extended-Phase-Shift Control. *IEEE Trans. Ind. Appl.* **2018**, *54*, 3483–3496. [CrossRef]
32. Li, S.; Yuan, X.; Wang, Z.; Wang, K.; Zhang, Y.; Wu, X. A Unified Optimal Modulation Strategy for DAB Converters to Tradeoff the Backflow Power Reduction and All ZVS in the Full Operating Range. *IEEE J. Emerg. Sel. Top. Power Electron.* **2023**, *11*, 5701–5723. [CrossRef]
33. Krismer, F.; Kolar, J.W. Closed Form Solution for Minimum Conduction Loss Modulation of DAB Converters. *IEEE Trans. Power Electron.* **2012**, *27*, 174–188. [CrossRef]
34. Everts, J.; Krismer, F.; Van den Keybus, J.; Driesen, J.; Kolar, J.W. Optimal ZVS Modulation of Single-Phase Single-Stage Bidirectional DAB AC-DC Converters. *IEEE Trans. Power Electron.* **2014**, *29*, 3954–3970. [CrossRef]
35. Tong, A.; Hang, L.; Li, G.; Jiang, X.; Gao, S. Modeling and Analysis of a Dual-Active-Bridge-Isolated Bidirectional DC/DC Converter to Minimize RMS Current With Whole Operating Range. *IEEE Trans. Power Electron.* **2018**, *33*, 5302–5316. [CrossRef]
36. Li, J.; Luo, Q.; Mou, D.; Wei, Y.; Sun, P.; Du, X. A Hybrid Five-Variable Modulation Scheme for Dual-Active-Bridge Converter With Minimal RMS Current. *IEEE Trans. Ind. Electron.* **2022**, *69*, 336–346. [CrossRef]
37. Xie, H.; Cai, F.; Jiang, J.; Li, D. A Fundamental Harmonic Analysis-Based Optimized Scheme for DAB Converters With Lower RMS Current and Wider ZVS Range. *IEEE Trans. Transp. Electrif.* **2023**, *9*, 4045–4058. [CrossRef]
38. Li, X.; Zhang, X.; Lin, F.; Sun, C.; Mao, K. Artificial-Intelligence-Based Hybrid Extended Phase Shift Modulation for the Dual Active Bridge Converter With Full ZVS Range and Optimal Efficiency. *IEEE J. Emerg. Sel. Top. Power Electron.* **2023**, *11*, 5569–5581. [CrossRef]

39. Zhao, B.; Song, Q.; Liu, W. Efficiency Characterization and Optimization of Isolated Bidirectional DC-DC Converter Based on Dual-Phase-Shift Control for DC Distribution Application. *IEEE Trans. Power Electron.* **2013**, *28*, 1711–1727. [CrossRef]
40. Liu, X.; Zhu, Z.Q.; Stone, D.A.; Foster, M.P.; Chu, W.Q.; Urquhart, I.; Greenough, J. Novel Dual-Phase-Shift Control With Bidirectional Inner Phase Shifts for a Dual-Active-Bridge Converter Having Low Surge Current and Stable Power Control. *IEEE Trans. Power Electron.* **2017**, *32*, 4095–4106. [CrossRef]
41. Yan, Y.; Luo, Q.; Luo, T.; Li, J.; Xiong, J.; Xiao, H. Current Stress Optimized Strategy for the Dual-Active-Bridge Converter With Triple-Phase-Shift and Variable DC-Link Voltage Control. *IEEE Trans. Power Electron.* **2025**, *40*, 8344–8355. [CrossRef]
42. Zhang, J.; Sha, D.; Song, K. Single-Phase Single-Stage Bidirectional DAB AC-DC Converter With Extended ZVS Range and High Efficiency. *IEEE Trans. Power Electron.* **2023**, *38*, 3803–3811. [CrossRef]
43. Cui, L.; Zhang, Y.; Wang, X.; Zhang, D. An Enhanced Integrated Optimization Strategy for Wide ZVS Operation and Reduced Current Stress Across the Full Load Range in DAB Converters. *Appl. Sci.* **2025**, *15*, 7413. [CrossRef]

MDPI AG
Grosspeteranlage 5
4052 Basel
Switzerland
Tel.: +41 61 683 77 34

Electronics Editorial Office
E-mail: electronics@mdpi.com
www.mdpi.com/journal/electronics

www.ingramcontent.com/pod-product-compliance
Lightning Source LLC
Chambersburg PA
CBHW061625210926
51850CB00009B/2437

Molecular Insights into the Developmental Origins of Health and Disease